7th and 8th grade levels

Hands-On Geometry

helping teachers
add classroom FUN
for learners

ACTIVITIES ON EARLY GEOMETRY
WITH CROSSNUMBER PUZZLES,
GEO-NOPOLY, PICTURE PUZZLE,
MINI-EXPERIMENTS, CONSTRUCTION
CLASSWORK, RIDDLES, MYSTERY
QUESTIONS AND DISCOVERY GAMES

WITH OPTIONAL ENRICHMENT EXERCISES
FOR SKILLBUILDING AND ENHANCING RECALL

plus solutions and answers

CESAR G. QUEYQUEP, Ph.D.

Department of Engineering
Purdue University Calumet
Hammond, Indiana, USA

Department of Mathematics
Bishop Noll Institute
Hammond, Indiana, USA

TRAFFORD
PUBLISHING

CANADA * US * UK * IREL;AND

Acknowledgement

The author acknowledges with thanks the valuable help of ... Adriana Cabrera-Cleves – publishing consultant, Connie McCann – pre-press technician, Veronica Dimofski – author services representative, Jon Poole – author services manager, and Owen Lett – pre-press technician ... (of Trafford Publishing) in the production of this book.

Cover: THE LOCKHEED SR-71 BLACKBIRD
Throughout its career (1973- 1997) of 24 years, the Blackbird remained the world's fastest and highest-flying plane, setting a world record of 2,193 MPH and altitude of 85,000 feet. It flew from LA to London, 5,645 miles, in just 3 hours and 47 minutes.

Most of its components are in the shape of circles, triangles, and trapezoids. Geometry plays a major role in the design of sport, commercial and military aircraft.

Note for Librarians: A cataloguing record for this book is available from Library and Archives Canada at www.collectionscanada.ca/amicus/index-e.html
ISBN 1-4120-9175-6

PUBLISHING™
Offices in Canada, USA, Ireland and UK

Book sales for North America and international:
Trafford Publishing, 6E–2333 Government St.,
Victoria, BC V8T 4P4 CANADA
phone 250 383 6864 (toll-free 1 888 232 4444)
fax 250 383 6804; email to orders@trafford.com
Book sales in Europe:
Trafford Publishing (UK) Limited, 9 Park End Street, 2nd Floor
Oxford, UK OX1 1HH UNITED KINGDOM
phone 44 (0)1865 722 113 (local rate 0845 230 9601)
facsimile 44 (0)1865 722 868; info.uk@trafford.com
Order online at:
trafford.com/06-0929

10 9 8 7 6 5 4 3 2

T o ...

Caroline 1, Lili, Rachel, Rebecca, and Lukas

.... for whose tomorrows,
this be a guiding light.

About the author

Dr. Cesar G. Queyquep earned his aeronautical engineering degree (top 10) from Feati University, his MS degree in mathematics education from Purdue University (as an NSF scholar), and his doctorate degree in education, summa cum laude, from Madison University. He was valedictorian of his graduating class at the Urdaneta Provincial East High School, now the Urdaneta City National High School.

Dr. Queyquep served on the faculty of the Department of Engineering of Purdue University Calumet (Northwest Indiana campus) as professor of Computer-Aided Design, Engineering Drawing, and Engineering Design. While connected with Purdue, he was also on the faculty of the Department of Mathematics at nearby Bishop Noll Institute.

At Bishop Noll Institute, he taught honors geometry, physics, algebra 1 and 2, industrial design, and mechanical drawing. He was Faculty Sponsor for the Math Club for three decades, Director of the Annual Mathematics Tournament for the Diocese of Gary and neighboring Chicagoland schools. He was also the Faculty Sponsor and Coach of the Bishop Noll JETS Academic Team which won three successive First Place Trophies in the Annual JETS Academic Tournaments at Valparaiso University and Purdue University Calumet. He was inducted into the Bishop Noll Institute Hall of Honor for outstanding service in the field of education to the Bishop Noll Community.

He authored two books published by the Kendall-Hunt Publishing Co., namely: "The Essential AutoCAD" (for engineering students in computer-aided design) and "The Essential Engineering Graphics Concepts" for engineering students. Both books were used at Purdue University Calumet, Ivy Tech State College Northwest, Bishop Noll Institute and other area schools, He was also a contributing author (he wrote the Research Project Report on "The Mathematics of Flight") in the Student Merit Awards book published by the National Council of Teachers of Mathematics in Reston, Virginia.

He taught college physics as a graduate teaching assistant at Western Michigan University. At Feati University, his alma mater, he was an assistant professor teaching aerodynamics, engineering physics, and theory of flight to aircraft maintenance, engineering and flying students, and later became Head of the Department of Aeronautical Engineering.

The activities in this book start with the very basic areas of geometry and include other related concepts that are expected to be within the reach of students in the 7th and 8th grade levels.

POINT A

This book recognizes the very significant starting point of learning geometry which deals with the basics, fundamentals, and elements upon which all other subsequent facts, concepts, and principles are based. The undefinables in geometry such as point, line, and plane provide the logical starting point, Point A, in presenting the exciting subject of geometry that it has been in civilizations past to the fast-paced technological world of our times.

OBJECTIVES / LEARNING TOOLS

The objectives describe the end results expected from each activity and provide the teacher and the learner with a clear idea of how to relate past knowledge to the understanding of concepts and principles to the task ahead. Knowing the activity goals enables the students to dig out the references essential and helpful to the learning process.

SUGGESTIONS FOR TEACHING

These suggestions are aimed at reminding teachers about the background needed by them for effectively presenting the lesson that day. They will then be ready to spend a few minutes of review to prepare the class for the activity.

MINI-EXPERIMENTS

These are small-scale gathering data and measurement types of work that will introduce the younger minds into the more serious tasks of research and discovery that lie ahead in their future careers. Knowing how to gather data and evaluate them into a useful piece of information strengthens their foundation for greater achievements in various areas of knowledge

PREPARE AHEAD

A successful presentation of the lesson or activity from day to day needs a thorough preparation. Teachers want to know what materials are needed and devices to be prepared in advance so as to effectively present the activity within a particular time module.

OPTIONAL ENRICHMENT EXERCISES

Learning skills can be built with simple follow-up exercises. These problems provide th means to build skills and enhance the power of recall.

EXAMPLES

Illustrative examples help the class understand concepts and principles better and give them confidence in what to do. Examples are shown in the activities to provide direction in problem solving.

SOLUTIONS AND ANSWERS

It is hoped that the solutions and answers at the back of the book would provide a helpful reference to the teacher and the learner. Illustrations and diagrams in the solutions both in the lessons and answers have been provided for clearer understanding of the givens, unknowns, and the desired results.

The crossnumber puzzles, picture puzzles, geo-nopoly, mini-experiments, construction exercises, and discovery games hopefully would provide adequate activity variation from one day to the next, and hopefully also, the funny nature of the activities and the extensive humor injected into the exercises would generate more student interest and greater response.

—— Cesar G. Queyquep, Ph.D.

NCTM Alignment

Most of the activities in this book deal with geometric shapes in one, two, or three dimensions. Points are one-dimensional, plane areas and surfaces are two-dimensional, and volumes which involve not only width and height, but also depth, are three-dimensional

The problems in the activities are about the various shapes that occur in nature and also in the imagination. They interrelate the one, two, and three-dimensional shapes and other geometric forms to each other, thus helping students see how these relationships affect our daily lives and influence the workplace and the physical environment.

Overview

Land of the pilgrims' pride

(The beginning concepts of geometry). Each member of the class answers questions about the concepts of point, line, segment, ray, and plane and in the process discovers the required solution to a given problem.

Race cars on the circle speedway

(Angle measurement). Groups of 4. Each group member is a race car driver, moving through angle measures around the track. First to make at least 360 degrees wins.

Trek to the stars

(Polygons). Groups of 4. Voyagers scan space for various polygon shapes among debris, meteorite chips, fallen satellite splinters, star dust, and other pieces of matter in the space beyond the earth. The drawer of the biggest polygon wins.

Better luck with horseshoes

(Building triangles). Groups of 4. The first group member to draw three required triangles wins.

Wheel of Pythagorean triples

(Right triangles). Groups of 4. Group members get two sides of a triangle by spinning a wheel, then figure out the length of the third side by using the Pythagorean Theorem.

The great triangle mystery

(Sum of the angles in a triangle). A mini-experiment to determine why the angles in a triangle always add up to 180 degrees.

Quadrilateral geo-nopoly

(Area of quadrilaterals). Groups of 4. Group members pretend they are realtors and play monopoly acquiring real estate. "Richest" realtor wins.

What really is a golden rectangle?

(Rectangles, golden rectangle). Individual. The whole class solves for areas of rectangles, search for golden rectangles and determine their areas.

How much area does a triangle have?

(Areas of triangles, using base and height). Individual. Determining the area of a triangle from its height (altitude) and its base.

Amazing area formula for triangles

(Area of a triangle, given 3 sides). Individual. The class has to figure out areas of triangles, with calculators, given the lengths of 3 sides.

The longest side of a triangle

(Pythagorean Theorem and right triangles). Groups of 4. Determining the longest side when the lengths of the legs of a right triangle are given.

No other triangles like these

(Special right triangles). Individual. Given, one side of a right triangle, the other sides are determined in various problems presented in a cross-number puzzle.

Overview

Always 18 on a Line

(Quadrilaterals). Individual. Solving for the areas of various quadrilaterals, given the base and the altitude, then arranging the areas so that thier sum horizontally, vertically, or diagonally is always 18.

Triangles in a bridge

Construction with triangles). Groups of 4. Groups build bridges from soda straw using triangular shapes. Gradually increasing loads are placed on the bridges until they collapse. The strongest bridge wins.

Rediscovering pi

(Circles and pi). Groups of 4. A mini-experiment on obtaining the value of pi. The pi value nearest to 3.1416 wins.

Around the house in circles

(Circumference and area of a circle). Individual. The class solves problems on the circumference and area of a circle while doing a picture puzzle.

Circles from triangles

(Areas of circles). Individual. Circle area problems presented in a crossnumber puzzle.

Circles on the target board

(Circular area strips). Groups of 4. Group members solve for areas of circular strips on a target board.

A hat for magicians and clowns

(Cones). Individual. Exercises on the volume of a cone, given the base diameter and its altitude, provide the solution to a puzzle.

Robin Hood could have drawn this with his arrows

(Coordinates of Points). Individual. The class draws figures by connecting points with given coordinates or ordered pairs).

A hat for "Mardi Gras"

(Constructing cones). Individual. Given the base diameter or radius, the altitude, and the slant height, students construct a cone much like a Halloween hat.

The longest flight

(Rectangular prism, cylinder, cone, and cube). Groups of 4. Students pretend they are on flight missions. The ship with the most fuel flies longest.

Why this turkey struts across the street?

(Symmetry). Individual. Students locate image points of given pre-image points to complete a symmetric figure and solve a given puzzle.

Sharks play what game in the ocean?

(Ordered pairs, coordinates of points). Individual. The class identifies the coordinates of points on an object to solve the puzzle about sharks.

Can you build a sphere?

(Spherical surface area). Groups of 4. A construction activity to find the area of a ball or sphere.

Table of contents

Learning Goal

The class will learn more about the beginning concepts of geometry, namely: point, line, segment, ray, and plane.

What was the boat that brought the pilgrims from England to Plymouth Rock in New England?

The solution to this problem may be found in the answers to the questions that follow. The following examples show how to identify and name various figures.

Teaching / Review notes

Review the class on the beginning concepts of geometry such as those shown below:

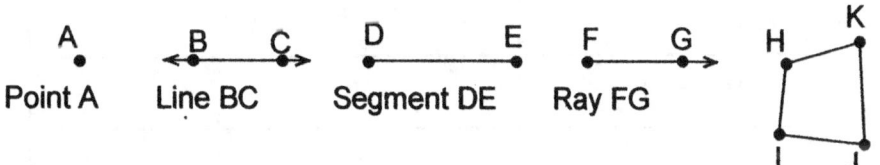

POINT. It has no definite size but is usually represented by a dot. It indicates location.

LINE. Formed by many points extending endlessly in opposite directions. It has no endpoints.

SEGMENT: A portion of a line, has two endpoints.

RAY. A portion of a line, has one endpoint and extends infinitely from that point to a given direction.

PLANE. Represented by any flat surface such as a book cover, window pane, or table top.

EXAMPLES

Connect the points shown below. Name the figures formed.

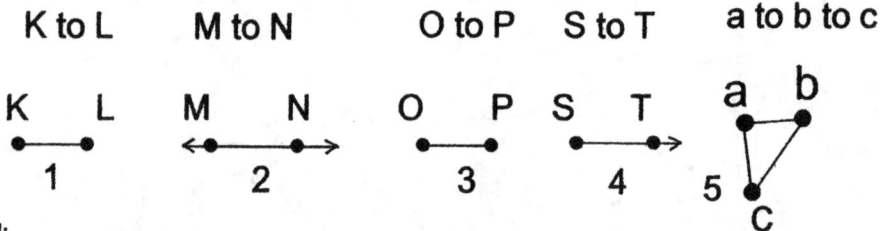

K to L   M to N   O to P   S to T   a to b to c

SOLUTION:
1. segment KL 2. line MN   3. segment OP 4. ray ST   5. plane abc

Directions

1. Refer to the map on the preceding page. Connect the points indicated in the left hand column of the table below. Write your answers on the blanks provided in the right-hand column.

POINTS TO CONNECT	WHAT KIND OF FIGURE IS FORMED? (Line, segment, ray or plane) Name the figure properly.
1. R to U	1.
2. E to J	2.
3. W to P	3.
4. O to X	4.
5. L to E	5.
6. F to S	6.
7. Y to T	7.
8. A to G	8.
9. M to Z	9.

2. Circle the first capital letter in the name of the indicated figure.
   Example: If the answer to a question is segment AB, place a circle around the letter A. If the answer to a question is CD, place a circle around the letter C.

3. Read all the circled letters IN REVERSE (BACKWARD) to get the desired answer.

Practice / Enrichment Exercises

Identify and name the figures shown below.

Figures	Names
1. E ● —— ● F	1. _____
2. K ● ——→ L	2. _____
3. S ● —— ● T	3. _____
4. ←● R ● V→	4. _____
5. L ●	5. _____
6. A B / D C	6. _____

```
←————●————●————●————●————●————●————●————●————→
      W     Y     O     A     B     C     D     E
    -10    -5     0     5    10    15    20    25
```

If numbers are assigned to equally-spaced points on a line like the one shown above, such a line is called a number line.

Can you solve for the lengths of the segments listed below? Indicate their lengths on the blanks provided at the right.

FIGURE	LENGTH	FIGURE	LENGTH
7. OC	_____	12. YB	_____
8. AE	_____	13. YO	_____
9. AC	_____	14 YC	_____
10. BE	_____	15. WD	_____
11. OE	_____		

16. If the left endpoint X of segment XY is marked -12 and XY is 30 inches long, the right endpoint Y should be marked what?

_____

Objective

Students will learn more about angles and how to measure angle size with the protractor.

Teaching suggestions

Ignite the enthusiasm of the class, explain to them that for today they make-believe that they are race car drivers on the speedway. The first one to complete the circular course WINS.

They imagine themselves as speedway drivers - wanna-be stars and raceway legends like Bobby Allison, one of the most successful drivers in American history: Dale Earhnhardt from North Carolina who won 74 NASCAR races and 7 championships; Jeff Gordon, youngest champion in NASCAR'S modern era; Alan Kulwicki, who won 5 Winston Cup Series races in his career; David Pearson, the "Silver Fox" for his winning style on the speedway; Richard Petty, who won 200 races in his 35-year career and won 7 Daytona 500's; Rusty Wallace, who won 40 races as well as the 1989 championship; Darrell Winthrop who won 84 races and 3 championships along the way; Dale Yarborough who won 83 races and ranks 5th in NASCAR's all-time wins. ; and many more.

Refresh the class on the meaning of angle, the use of the protractor, and the various kinds of angles. Review also adjacent, complementary and supplementary angles, vertical angles, triangles, and why the sum of the angles in any triangle and why the sum of the angles in any triangle is always equal to 180 degrees.

Teaching notes

An angle is the figure formed by two rays, lines, or segments, that have a common endpoint called the vertex of the angle. The angle in the figure is ∠ABC, with B as its vertex.

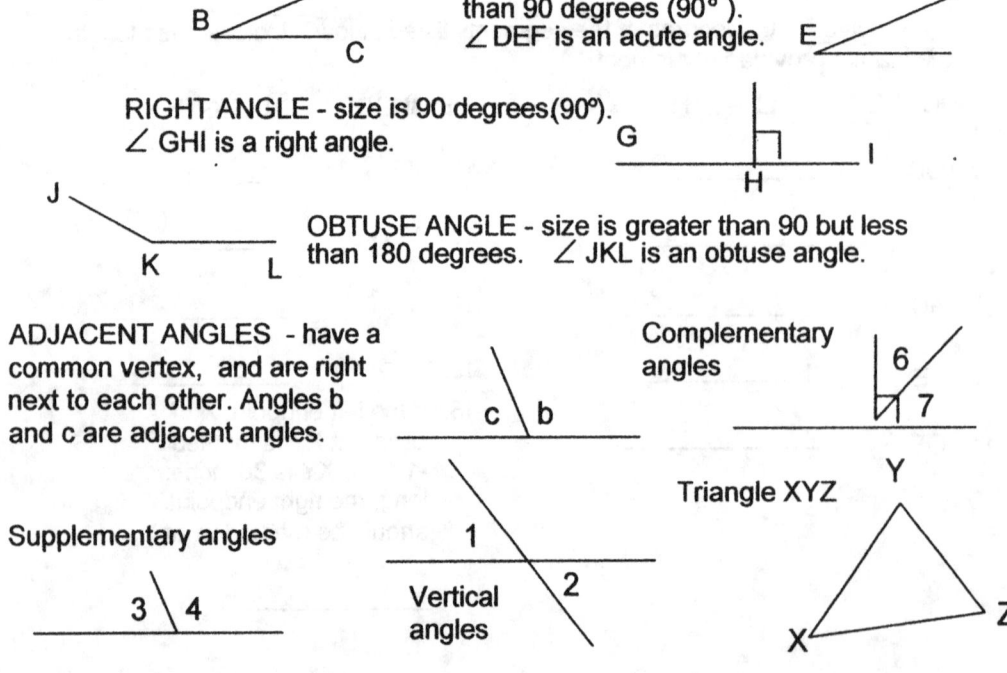

- ACUTE ANGLE - size is less than 90 degrees (90°). ∠DEF is an acute angle.
- RIGHT ANGLE - size is 90 degrees(90°). ∠ GHI is a right angle.
- OBTUSE ANGLE - size is greater than 90 but less than 180 degrees. ∠ JKL is an obtuse angle.
- ADJACENT ANGLES - have a common vertex, and are right next to each other. Angles b and c are adjacent angles.
- Complementary angles
- Supplementary angles
- Vertical angles
- Triangle XYZ

Prepare ahead

Using index cards or manila folder, cut out a big triangle 2" on each side.

Fold the triangle shown below at BV, VC, and BC, then use scotch tape to assemble the pyramid at the right below. Write 10, 20, 30, and 40 on the four faces as shown.

Make 6 of these pyramids, one for each group of 4. Build more if needed.

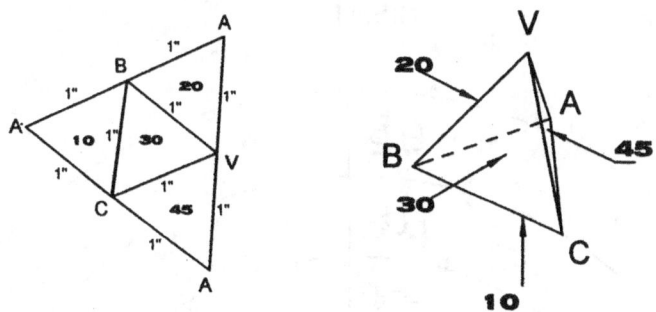

Photocopies of the Speedway on the next page for each member of the class.
Protractors for each group.

Directions

(Before the game starts, students should review how to use the protractor, and practice a few measurements. Walk around the groups to provide help where needed.)
Divide the class into groups of 4.
Each member of a group gets a photocopy of the race track.
Each player starts on the start line.
Player 1 "rolls" the pyramid as he would roll a pair of dice.

If the number hidden at the base of the pyramid is 20, he draws a line from center O, the line OA forming 20° with the start line.

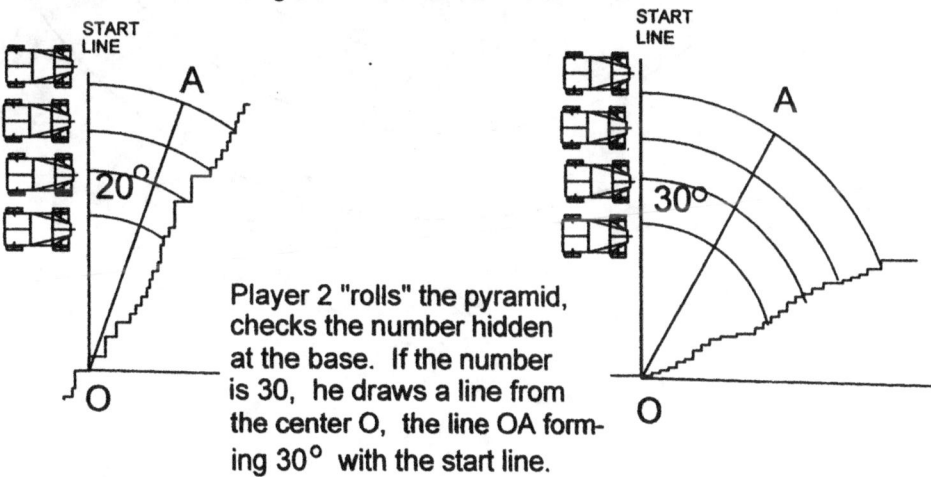

Player 2 "rolls" the pyramid, checks the number hidden at the base. If the number is 30, he draws a line from the center O, the line OA forming 30° with the start line.

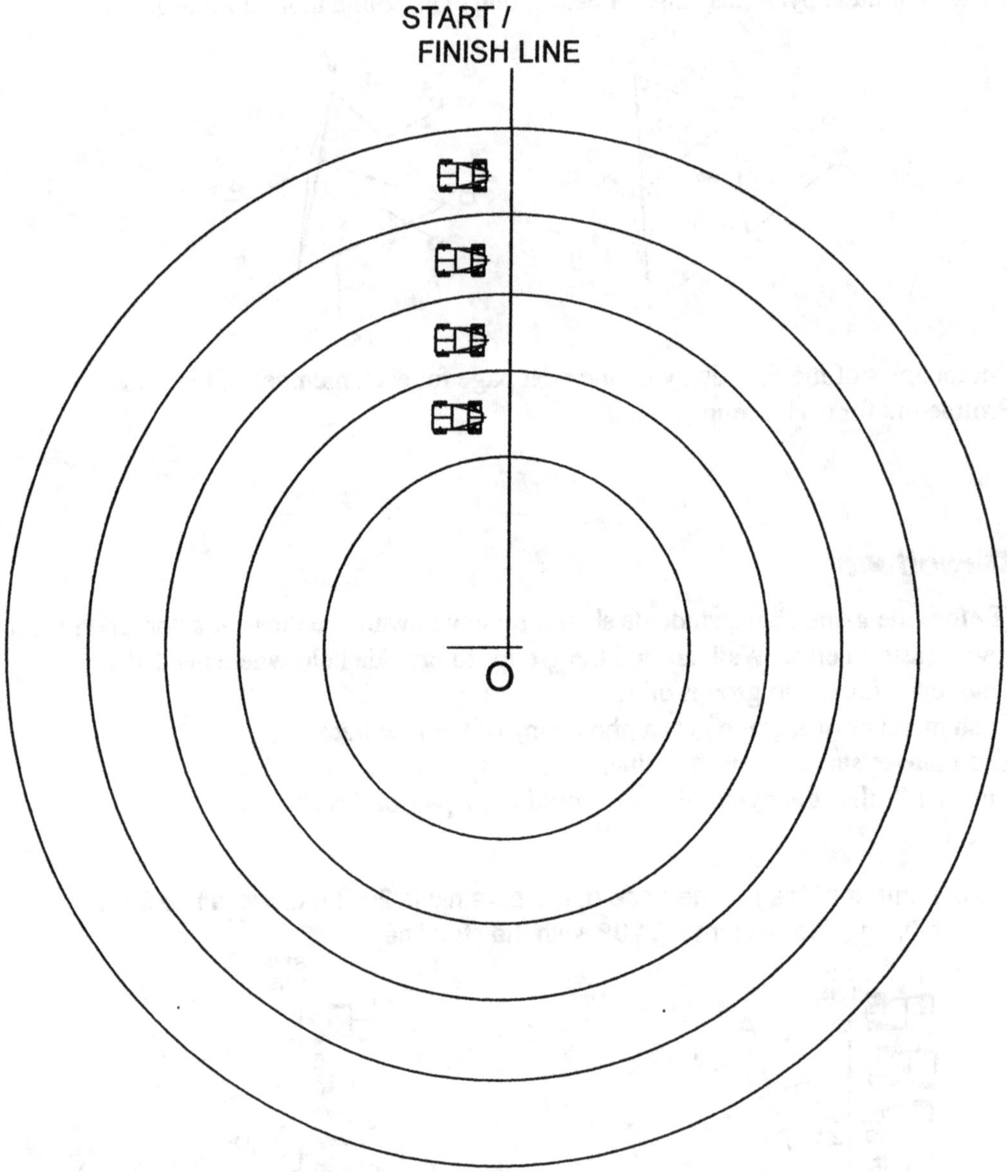

START /
FINISH LINE

O

Example 2

Player 3 rolls the pyramid, checks the number hidden at the base, and finds the number 30. He draws a line from the center O, the line forming an angle of 30 degrees with the start line as shown in the figure below.

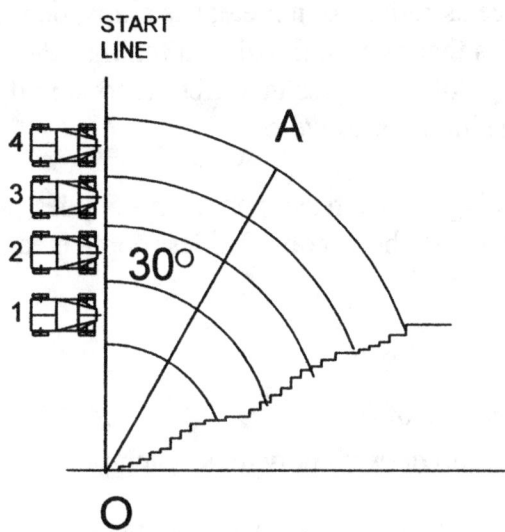

Optional enrichment exercises

ON YOUR OWN
TRY THESE SKILLBUILDERS AND RECALL ENHANCERS

1. An angle is the figure formed by two rays or two segments that have a common

   _____.

2. A (an) _____ angle has a size of less than 90 degrees.

3. A _____ is a plane figure with three sides.

4. If angle ABC is a right angle, it has a size of _____ degrees.

5. An angle with a size of more than 90 degrees is called a (an) _____ angle.

Learning objective

Students will learn more about the properties of polygons, their sides and angles, and their relationship to the circle.

Teaching notes

Bring the class with you on a Field Trip to the stars, into space "where no man has gone before". Pretend that you are all on a voyage to see the geometry of polygons in outer space as reflected in meteorite chips, debris from space satellites, and mini-asteroids that float in the distant realms beyond our solar system. It is a fantasy expedition of your class in a lifetime to identify various shapes of polygons, if any, in the heavens above us.

Review the class on polygons, such as: pentagon - 5 sides(penta means 5), hexagon - 6 sides(hexa means 6), heptagon - 7 sides (hepta means 7), etc.

Directions

* The class works in groups of 4.
* Each member of a group draws 3 polygons, each polygon having sides of EQUAL lengths.
* No two members of each group should have polygons of the same number of sides. Each polygon should have at least 5 sides.
* Each group will have 12 polygon drawings.
* The group that draws the polygon with the MOST number of sides wins. In case of a tie, the teacher decides which is the better-drawn, better-looking polygon.

Examples:
_____

To construct:

Pentagon, no. of sides
n = 5 ; 360 /5 = 72°

Hexagon
n = 6; 360 / 6 = 60°

Octagon
n = 8; 360 /8 = 45°

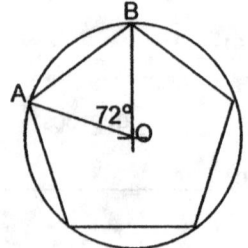
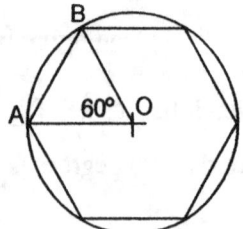
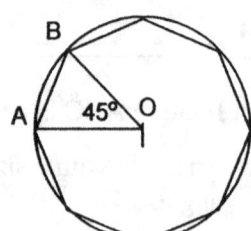

The winner writes on the board the information below about the polygon.

Length of one side _____

No. of sides _____   Size of angle AOB _____

He then asks the class:

For the correct name of the polygon. _____ `1`
  (1 point for correct answer as shown in the box at the right)

What happens if angle AOB is made equal to 0? _____ `3`
  (3 points for correct answer as shown in the box)

_____

If angle AOB is zero, what would figure AOB look like? _____ `9`

_____

What then would be the length of one side of the polygon? _____

_____ `18`

What would be the shape of triangle AOB? _____

_____ `3`

How many sides would that polygon have? _____

_____ `12`

How would the polygon look like at this point? _____

_____ `5`

(To check your answer to the last question, find the words spelled out by the 1st, 3rd, 9th, 18th, 3rd, 12th, and 5th letters of the alphabet. This is the MYSTERY POLYGON the class is looking for.)

Prepare ahead

Ruler for each group
Protractor for each group
Photocopy of page 2 of this activity for all members of the class
Typing paper for each member of the class (for drawing their polygons)

Optional enrichment exercises

ON YOUR OWN
TRY THESE SKILLBUILDERS AND RECALL ENHANCERS

A figure that has many sides is called a _____.

The (word) prefix poly means   a) three   b) few   c) five   d) many

A polygon with _____sides is called a regular polygon.   This

kind of a polygon can be drawn inside (inscribed) in a circle.

The angle AOB at the right is called a central
angle. The polygon has 6 sides, and also 6
central angles. The size of each central angle
is   a) 360   b) 60   c) 100   d) 360 / n,
where n = the number of sides.

A polygon with 8 sides is called a(an) _____ .

A decagon has _____ sides.

The measure (size) of the central angle of a decagon is _____

_____ degrees.

Learning goal

This activity clarifies the relationship between angles and triangles.

Teaching suggestions

Triangles can be isosceles, acute, scalene, equilateral, obtuse and right. Emphasize to the class that no matter what kind of a triangle it is, the sum of its three angles is always equal to 180 degrees.

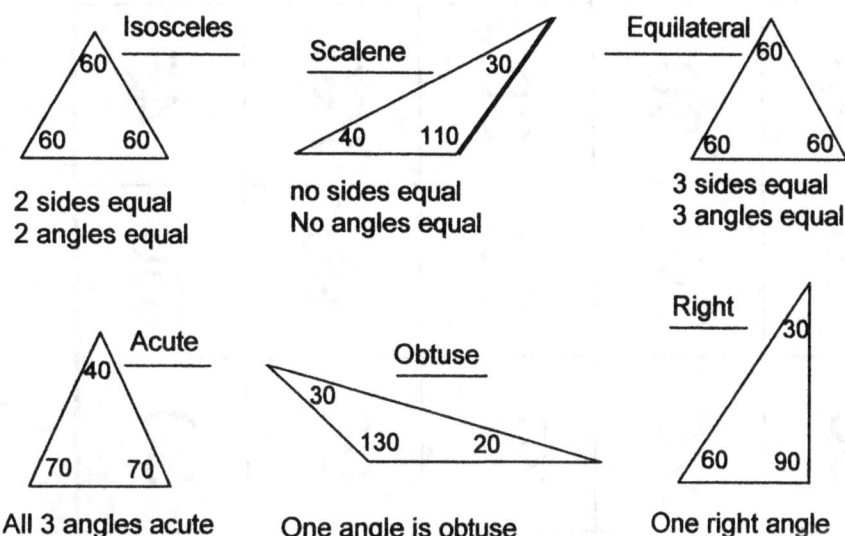

Directions

Divide the class into groups of 4.

Each group will need a) a horseshoe cut from thick cardboard or manila folder, refer to actual size and shape on the next page of this activity, and b) a Squares Grid as shown in the figure below. Refer also to the Squares Grid on the next page.

Example 1.

Each group member throws the horseshoe into the grid. If it lands on a 30° square, he draws a 30° angle on his paper. If, on his next throws on his next turns he gets 60 and 90 on the grid, he completes a 30-60-90 right triangle. He then draws a 30-60-90 triangle like the one shown at the right above.

HORSESHOE

30	90	30	60	45	120
30	120	75	30	60	30
60	30	75	90	30	45
75	120	90	90	120	60
30	90	30	60	45	30

SQUARES GRID

SQUARES GRID

120	45	60	30	90	30
30	60	30	75	120	30
45	30	90	75	30	60
60	120	90	90	120	75
30	45	60	30	90	30

HORSESHOE
(Actual Size)

* The squares grid should be on the floor.

* Players should be 3 ft away from the grid when they cast the horseshoe.

* For a throw to be VALID, the horseshoe must not touch any line, as shown on "75" at the right, on the grid.

Example 2.

In his turns, Player 2 gets 120, then 30, and then another 30. This gives him the angles for his triangle. It turns out that the triangle has two equal sides and is therefore an isosceles triangle.

The first in a group to draw three triangles shown on page one of this activity WINS and is the group winner.

The winners of all groups will be asked by the teacher to go to the board to answer the following problem.

Given: the figure below is a portion of a polygon inscribed in a circle.

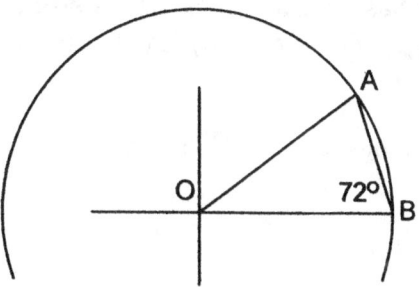

Figure out how many sides this polygon will have.

The first 3 to finish with correct answers will be the class winners.

Prepare ahead

Horseshoe, cut from thick cardboard or manila folder, for each group in class
Photocopy of Squares Grid, each taped to manila folder, for each group in class.
Protractor, ruler, and scotch tape for each group in class
Typing paper (for each student) on which to draw the required three triangles.

Optional enrichment exercises

ON YOUR OWN
TRY THESE SKILLBUILDERS AND RECALL ENHANCERS

1. If the angles of a triangle are $60°$-$65°$-$55°$. what kind of a triangle is it?

2. The sides of a triangle are 12", 15", and 16". What kind of a triangle is it?

3. Is it possible to have two right angles in a triangle? Explain.

4. A triangle has base angles (angles at its base) of $60°$ and $60°$. This is what kind of a triangle?

5. A triangle has angles of $30°$ -$60°$ -$90°$. What kind of a triangle is this?

6. A regular polygon has 12 sides. What is the size of one of its central angles?

7. The angles of a triangle are $50°$-$50°$-$80°$? What would you call such a triangle?

8. Two angles of a triangle are $110°$ and $50°$. What kind of a triangle is this?

Objective

How to determine the length of the sides of a right triangle.

Directions

* Explain to the class that the longest side of a right triangle is called hypotenuse, and that the two sides that are perpendicular to each other are the legs.

* Discuss with them the Pythagorean Theorem which states that
$$c^2 = a^2 + b^2$$

* Emphasize that in the Pythagorean Theorem, a and b represent the legs of a right triangle and that c is the hypotenuse.

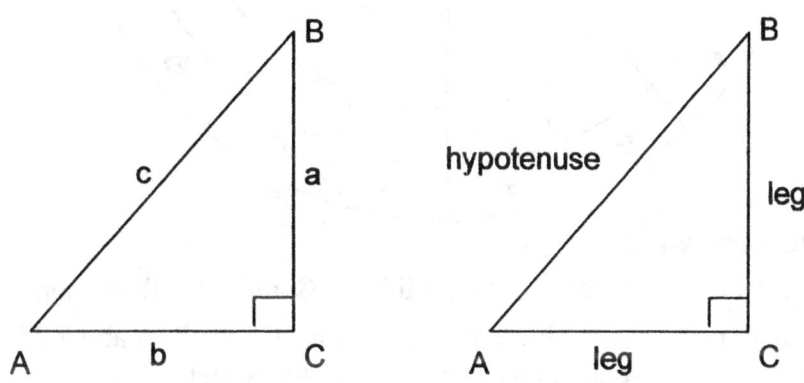

Example 1

In the table below, a = 4, b = 3, and c = 5. Observe that $a^2 = 16$, $b^2 = 9$, and $c^2 = 25$. Thus, 3, 4, and 5 are a Pythagorean Triple.

Example 2

The numbers 6, 8, and 10 are also a Pythagorean Triple. If a = 6, b = 8, and c = 10, note that $a^2 = 36$, $b^2 = 64$ and $c^2 = 100$. The numbers add up.

a	a^2	b	b^2	c	c^2
4	16	3	9	5	25
6	36	8	64	10	100

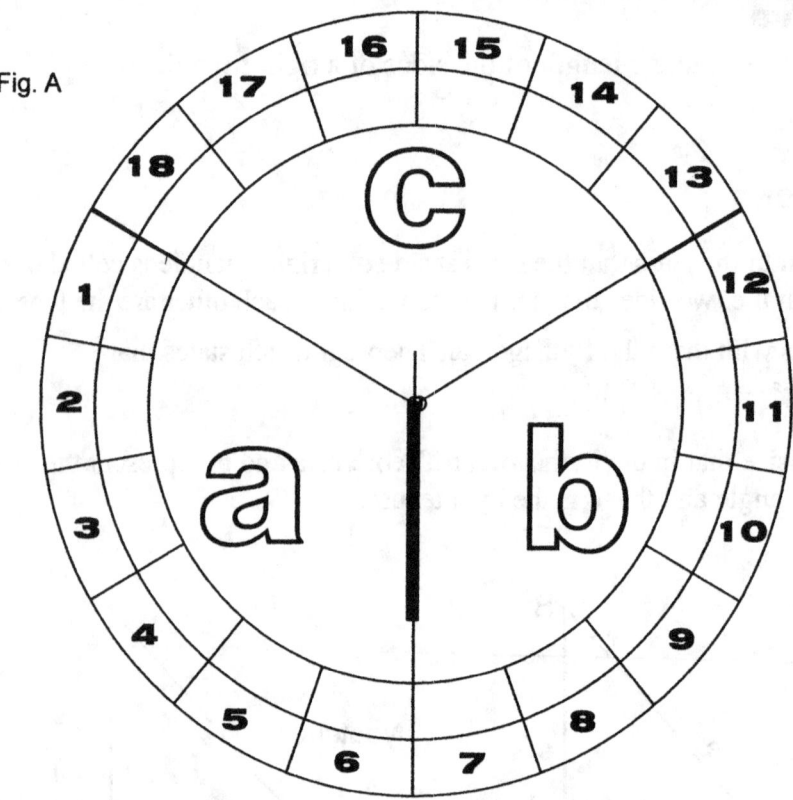

Fig. A

To create the wheel

1. Glue Fig. A on posterboard or manila folder. Cut it out with scissors .

2. Stick a paper clip through the center of the wheel, and allow about 2 inches length of the clip to stick out as a pointer over the wheel.

3. Tape the paper clip to a pencil, and there you have a spining wheel.

Playing the wheel

Each group member spins the wheel twice and reads two numbers for each turn. One could be an _a_ number, and the other a _b_ number; or one could be an _a_ number and the other a _c_ number. In either case, solve for the third number - which is either an _a,_ or a _b_ , or a _c_ ; if the number solved for is on the wheel, then there is a Pythagorean Triple.

Example 3

Player 1 gets a = 5 and c = 13. He records these readings in the table below. He records also values for $a^2 = 25$ and $c^2 = 169$.

Using the formula $\quad a^2 + b^2 = c^2$

$$25 + b^2 = 169$$
$$b^2 = 144, \quad \text{which yields } b = 12. \text{ (Use a calculator if needed.)}$$

a	a^2	b	b^2	c	c^2
5	25	12	144	13	169

Table 1   PYTHAGOREAN TRIPLES

Turn	a	a^2	b	b^2	c	c^2
1						
2						
3						
4						

The first group member to have 4 Pythagorean Triples is the winner of that group.

The first 3 groups to have 4 Pythagorean Triples are the class winners.

Prepare ahead

Photocopy of Fig. A(Wheel of Pythagorean Triples) and Table 1 (Pythagorean Triples) for the whole class.

Glue, scissors, posterboard / manila folder for each group

One calculator for every group

Optional enrichment exercises

ON YOUR OWN
TRY THESE SKILLBUILDERS AND RECALL ENHANCERS

1. Find c if a = 1.5" and b = 2".

2. A baseball diamond is 20 ft by 20 ft. How far is the third base from the home base? _____

3. The dimensions of a rectangle are l = 10" and w = 12"? How long is its diagonal? _____

4. A photograph is 8" long and 10" wide. How long is it from corner 1 to corner 3? _____

5. The shorter sides of a triangle are called its _____ .

6. The sum of the two small angles of a right triangle is how many degrees? _____

7. The longest side of a right triangle is always opposite the _____ angle.

8. The legs of a right triangle are 20" and 25" long. What is the length of the hypotenuse? _____

Objective

Students will solve the great triangle mystery - what is the sum of the angles of any triangle? Also, what is the sum of two adjacent angles whose other sides lie on a straight line.

Directions

Every member of the class participates in the experiment.

Part 1. Each student draws a straight line. At some point O close to the mid-point of the line, draw another line OM as shown below.

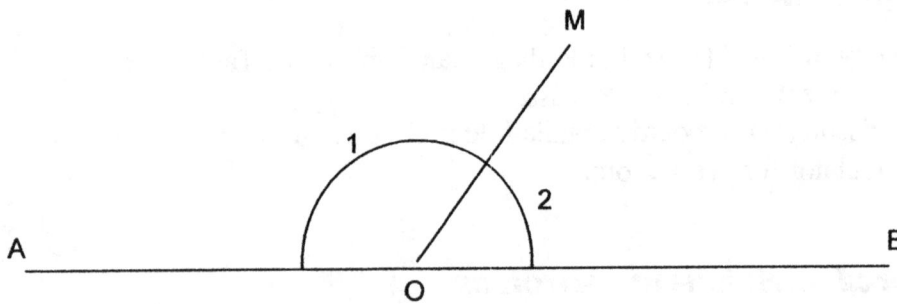

Part 2. Draw any triangle ABC, similar to the one shown below, on an index card. Cut out from the index card two more triangles, DEF and KLM both identical to triangle ABC.

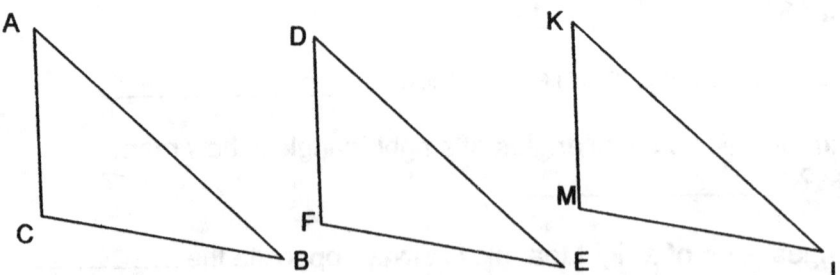

Arrange the three triangles ABC, DEF, and KLM as shown above, so that angles 1, 2, and 3 are adjacent to each other as shown below.

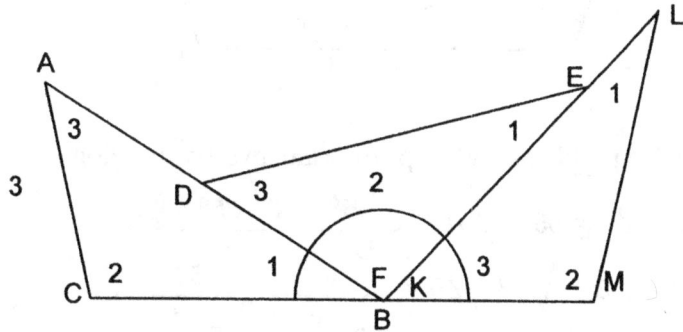

Use the protractor to measure the angles and record the data in the following table:

Triangle	$\angle 1$	$\angle 2$	$\angle 3$	$\angle 1 + \angle 2 + \angle 3$
ABC				
DEF				
KLM				

From the above figure and table, you may now answer the mystery question - Why do the three angles of the triangle always add up to 180 degrees?

The answers to the following problems will help solve the mystery question.

The three triangles at the right
have all their sides equal and
all their angles equal.

$\angle 1 = \angle 7 = \angle 9$

$\angle 4 = \angle 2 = \angle 8$

$\angle 5 = \angle 6 = \angle 3$

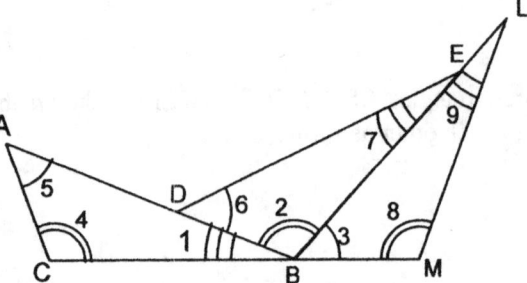

The three triangles at the right have all their sides equal and their angles equal.

∠1 = ∠7 = ∠9
∠4 = ∠2 = ∠8
∠5 = ∠6 = ∠3

The answers to the following problems will help solve the mystery question.

1. Given, ∠1 = 30°, ∠2 = 95°, ∠5 = $\dfrac{55°}{D}$, $\dfrac{65°}{I}$.

2. Given, ∠7 = 40°, ∠8 = 110°, ∠5 = $\dfrac{35°}{J}$, $\dfrac{30°}{H}$.

3. Given, ∠2 = 100°, ∠5 = 60°, ∠7 = $\dfrac{20°}{E}$, $\dfrac{25°}{K}$.

4. If ∠1 = 28°, ∠2 = 102°, ∠6 = $\dfrac{55°}{L}$, $\dfrac{50°}{G}$.

5. If ∠2 = 103°, ∠6 = 47°, ∠1 = $\dfrac{30°}{F}$, $\dfrac{35°}{M}$.

6. Given, ∠2 = 105°, ∠7 = 40°, ∠5 = $\dfrac{30°}{X}$, $\dfrac{35°}{S}$.

7. Given, ∠2 = 110°, ∠7 = 50°, ∠3 = $\dfrac{20°}{W}$, $\dfrac{30°}{Y}$.

8. If ∠8 = 108°, ∠9 = 40°, ∠6 = $\dfrac{35°}{Z}$, $\dfrac{32°}{V}$.

9. If ∠2 = 107°, ∠7 = 38°, ∠3 = $\dfrac{35°}{U}$, $\dfrac{33°}{A}$.

10. Given, ∠6 = 38°, ∠7 = 60°, ∠5 = $\dfrac{80°}{B}$, $\dfrac{82°}{T}$.

D. PLACED SIDE TO SIDE
H. WITH ALL
E. HAVING
G. A COMMON
F. VERTEX,

I. PLACED END TO END
J. WITHOUT ALL
K. ON THE SAME
L. VERTEX
M. POINT

S. THEIR OUTER SIDES
W. LIE
V. ON'A
U. STRAIGHT
T. LINE

X. THEIR OPPOSITE SIDES
Y. DO NOT LIE
Z. ABOVE
A. A HALF
B. CIRCLE.

Copy, on the lines below, the words or phrases that correspond to the letters beneath your answers to the problems above.

_____   _____   _____   _____

_____   _____   _____   _____

_____   _____

Prepare ahead

Photocopy of pages 1 and 2 of this activity for each member of the class
Protractor, ruler

Optional enrichment exercises

ON YOUR OWN
TRY THESE SKILLBUILDERS AND RECALL ENHANCERS

1. The sum of two adjacent angles, with a common vertex, whose outer sides form a straight line is _____ degrees.

2. The sum of two or more angles, with a common vertex, whose outer sides form a straight line is _____ degrees.

3. The sum of all the angles in any triangle is equal to _____ degrees.

4. If two right angles are adjacent to each other and have a common vertex, their outer sides would form a _____ line.

Learning Goal

The class will improve on the application of the formulas for the area of quadrilaterals such as the rectangle, square, rhombus, and trapezoid.

Teaching suggestions

Excite the class. Tell them that they are going to be realtors for the day. As realtors, they will play a game of Geo-nopoly, very much like the famous monopoly game, except that Geo-nopoly involves finding the areas of various quadrilaterals. Review the class on identifying the quadrilaterals, determining the perimeter and area of the rectangle, square, trapezoid, and rhombus.

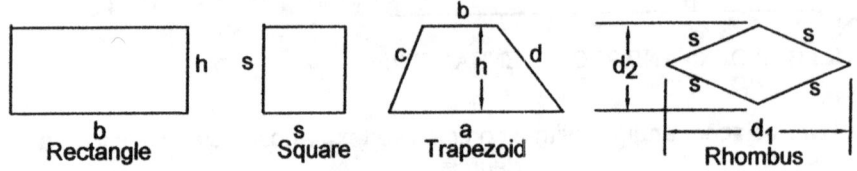

QUADRILATERAL	PERIMETER	AREA
Rectangle	2s + 2h	w x h
Square	4s	s x s
Trapezoid	a + b + c + d	1/2 (a + b) x h
Rhombus	4s	$1/2 \times d_1 \times d_2$

Example 1

Find the perimeter and area of rectangle ABCD at the right.

Solution:

P = 2w + 2h = 2 x 12 + 2 x 10 = 44 in.
A = w x h = 12 x 10 = 120 sq. in.

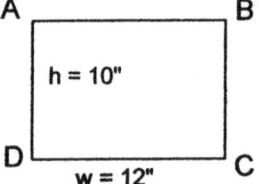

Example 2

Determine the perimeter and area of the rhombus KLMN at the right if d = 8" and d = 6".

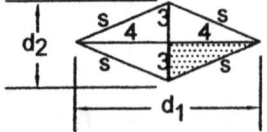

Solution:

The rhombus is made up of 4 right triangles. In the shaded triangle, the value of s can be found to be 5", by using the Pythagorean formula.
Hence, the perimeter P = 4s, or P = 4 x 5 = 20".

and $A = 1/2 \times d_1 \times d_2$

$= 1/2 \times 8 \times 6 = 24$ sq. in.

Part 1. Perimeters and Areas

Each correct answer is worth $25.

The total earnings of each player in Part 1 will be added to the total dollar sales in Part 2 of this game.
The player with the most money at the end of the game WINS.

Find the perimeter: ($25 for each correct answer. The teacher moves around the groups to check correctness of answers. No group credit for incorrect answers).

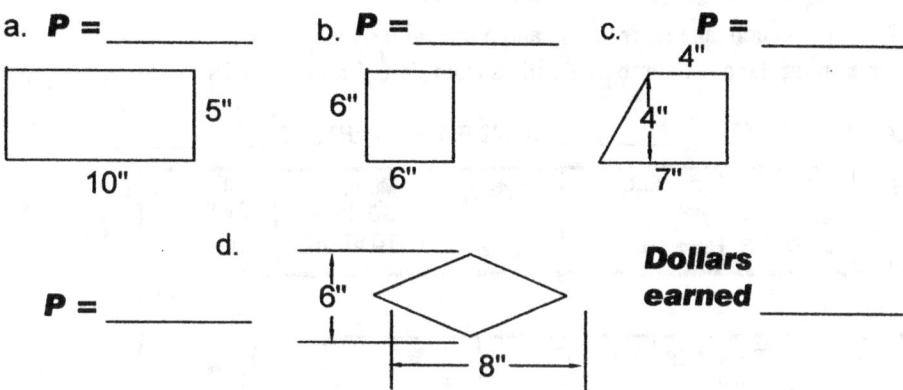

a. **P =** _____

b. **P =** _____

c. **P =** _____

d. **P =** _____

Dollars earned _____

Find the area: ($25 for each correct answer)

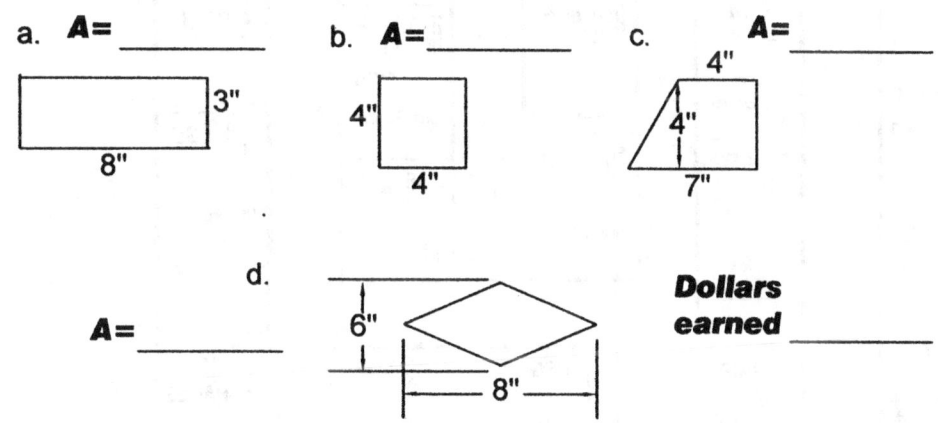

a. **A=** _____

b. **A=** _____

c. **A=** _____

d. **A=** _____

Dollars earned _____

Prepare in advance

Photocopy of pages 1 and 2 of this activity for each member of the class
A numbers cube (a die, like the die used in monopoly games) for each group

Part 2. Quadrilateral Geo-nopoly

DIRECTIONS

1. Divide the class into groups of 4. Each group member should have a copy of this page which shows the GEO-BOARD (travel route) below. The group should also get one die. (A die like the die used in monopoly games). To mark locations on the geo-board, a coin - a penny, nickel or dime - can be used.

2. Each player rolls the die to determine who starts first. The player who gets the highest number on the die starts first. Players take turns clockwise in rolling the die. If player 1 rolls the die and the die shows the number 3, he moves his coin up by 3 spaces, and comes to the space marked T. T stands for a TRAPEZOIDAL lot, and LGE means a large lot with an area of 7000 sq. ft. and a selling price of $7000. Player 1 then enters 7000 in the Individual Earnings Record on the next page.

3. Players 2, 3, and 4 each take their turns, and play similarly as player 1.

4. The player earning the most combined dollars from Parts 1 and 2 WINS.

DOLLARS EARNED IN PART 2 _____   TOTAL EARNED PARTS 1 & 2 _____

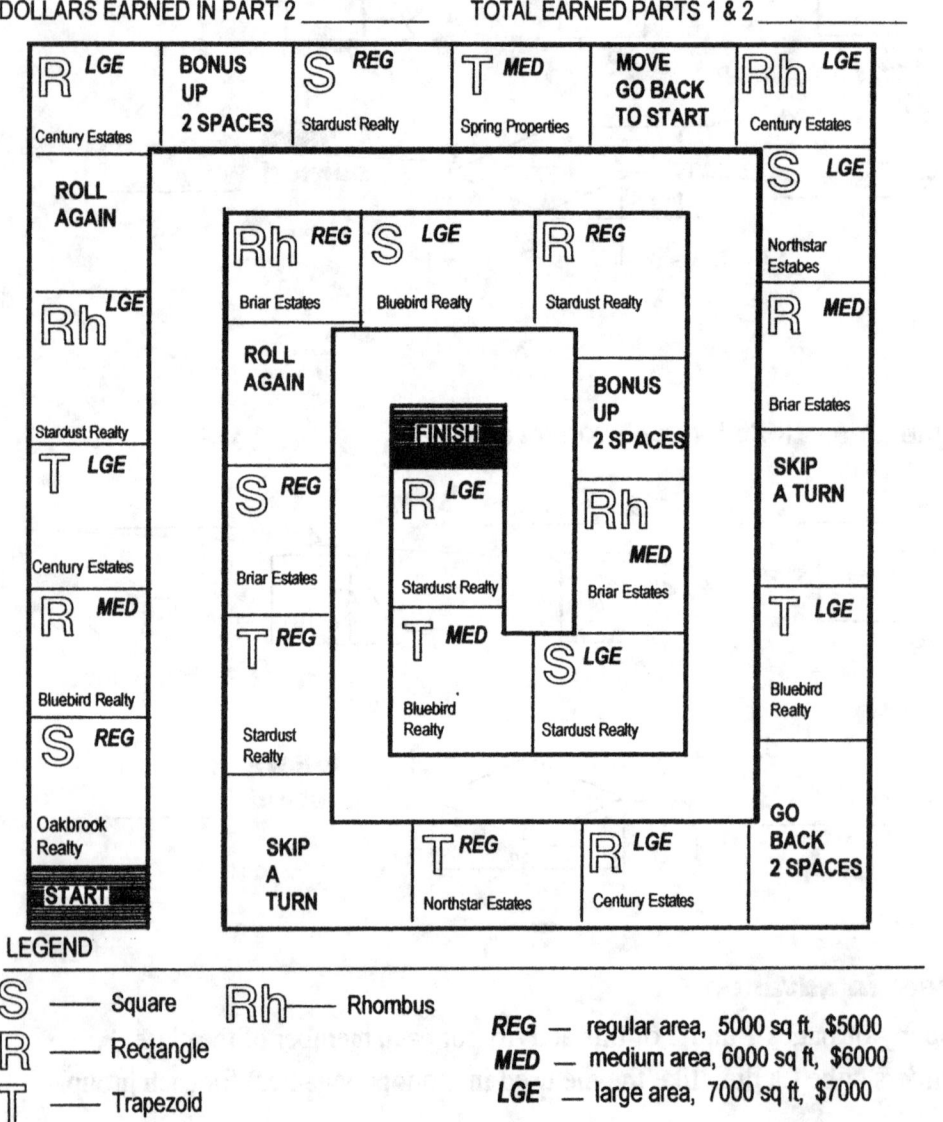

LEGEND

S — Square   Rh — Rhombus

R — Rectangle

T — Trapezoid

REG — regular area, 5000 sq ft, $5000
MED — medium area, 6000 sq ft, $6000
LGE — large area, 7000 sq ft, $7000

RECORD OF INDIVIDUAL EARNINGS

ROLL NO.	EARNINGS	RUNNING TOTAL	ROLL NO.	EARNINGS	RUNNING TOTAL
1			16		
2			17		
3			18		
4			19		
5			20		
6			21		
7			22		
8			23		
9			24		
10			25		
11			26		
12			27		
13			28		
14			29		
15			30		

Part 1. Points earned _____

Part 2. (After each roll, write earnings and running totals on the blanks above).
Points earned _____

GRAND TOTAL OF EARNINGS (Combine Parts 1 & 2 earnings)

_____

TO DETERMINE THE CLASS WINNER

1. The group member with the highest grant total points is the GROUP WINNER; and
2. The group winner with the highest grand total points is the CLASS WINNER.

Learning goal

Students will learn more about rectangles, the Golden Rectangle, and how their areas are determined.

Suggestions for teaching

The Greeks had a reason for calling this kind of a rectangle the Golden Rectangle - it has well-proportioned sides and is very pleasing to the eye.

The Golden Rectangle has this special property that when a square portion of it is removed, the dimensions of the remaining figure have the same ratio as the dimensions of the original rectangle. This ratio of its width w to its height h is $w/h = 1.62$.

For golden rectangle DEFG,
$$w/h = h/x = 1.62$$

Example 1

Given, rectangle DEFG, with w = 4.86 inches. What value of h would make DEFG a golden rectangle?

Solution: Since $w/h = 1.62$, $h = w/1.62$, hence $h = 4.86/2 = 3"$.

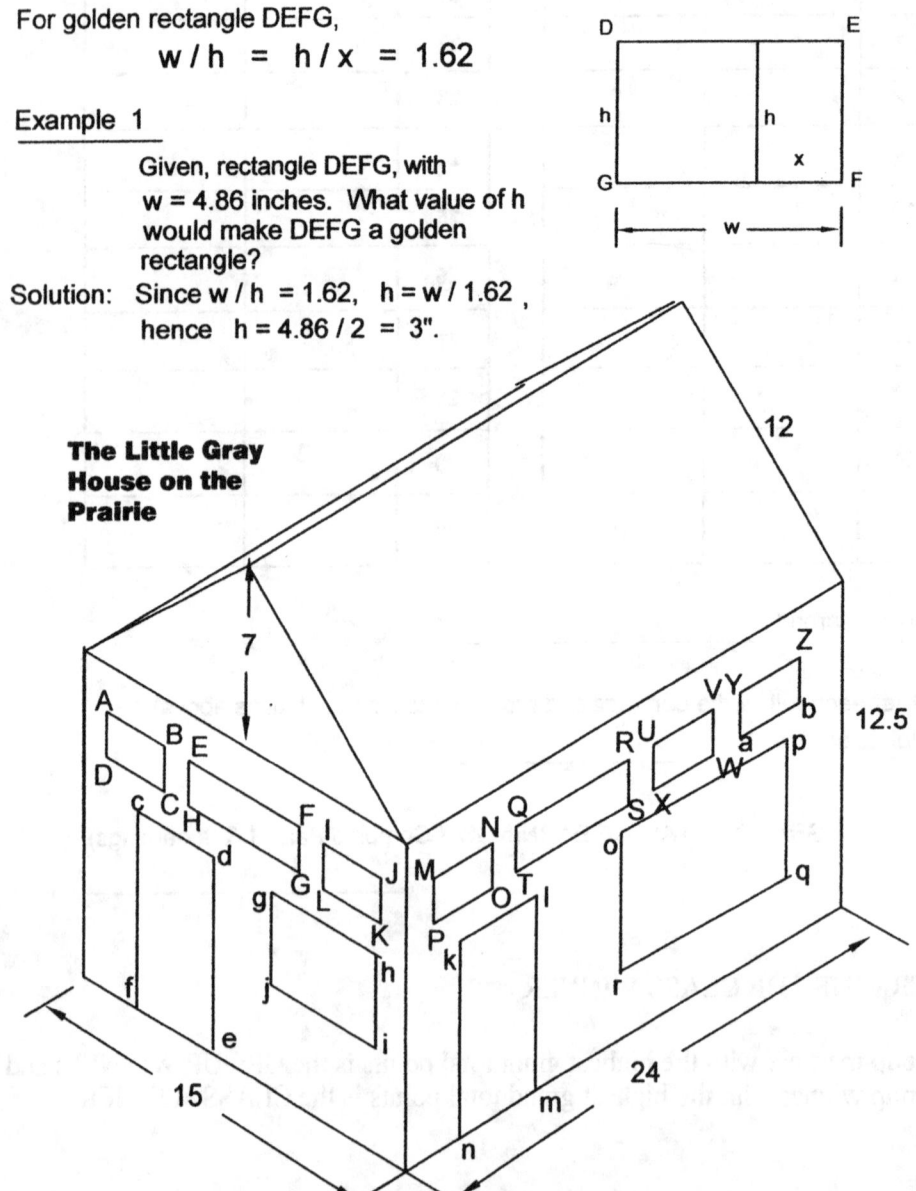

The Little Gray House on the Prairie

What really is a golden rectangle ?

Directions

Part 1.

Refer to the figure on the preceding page.

Determine whether or not each given rectangle is a golden rectangle. If it is, write "G" on the blank in the table below; if not, write "No".

Example 2

Divide w by h. In rectangle ABCD, w = 3.5 and h = 2.19.

w / h = 3.5 / 2.19 = 1.59 or 1.6 rounded to the nearest tenth. Write "G" on the first blank in the table as shown.

Note: If h > w , then instead of dividing w by h, divide h by w. If h / w = 1.6, the figure is a golden rectangle.

RECTANGLE	W	H	G / NG	RECTANGLE	W	H	G / NG
ABCD	3.5	2.19	G	YZab	3.5	2.19	____
EFGH	6	2.19	____	cdef	4.69	7.5	____
IJKL	3.5	2.19	____	ghij	6.25	3.9	____
MNOP	3.5	2.19	____	klmn	4.69	7.5	____
QRST	6.8	2.19	____	opqr	7.5	4.68	____
UVWX	3,5	2.19	____				

Part 2.

Find the area of rectangles
klmn and opqr.

Rectangle klmn, area = _____

Rectangle opqr, area = _____

Example 3

Find the area of a rectangle with w = 10 inches and h = 2.8 inches.

Solution:
　　　Area of a rectangle = w x h - 10 x 2.8 = 28 sq. in.

Prepare ahead

Photocopies of both pages of this activity for each member of the class
Calculators, if allowed, for quick division

Objective

To identify the base and altitude (height) of a triangle, and to determine its area.

Teaching suggestions

* Emphasize to the class that the altitude of a triangle and its base are perpendicular to each other;

* A triangle is a polygon of three sides;

* The formula for the area A of a triangle is $A = 1/2 \times \text{base} \times \text{height}$;

* The area must be expressed in square units such as square inches, square feet, square miles, square centimeters, etc.

The heavy lines are the base and altitude of the triangles shown below.

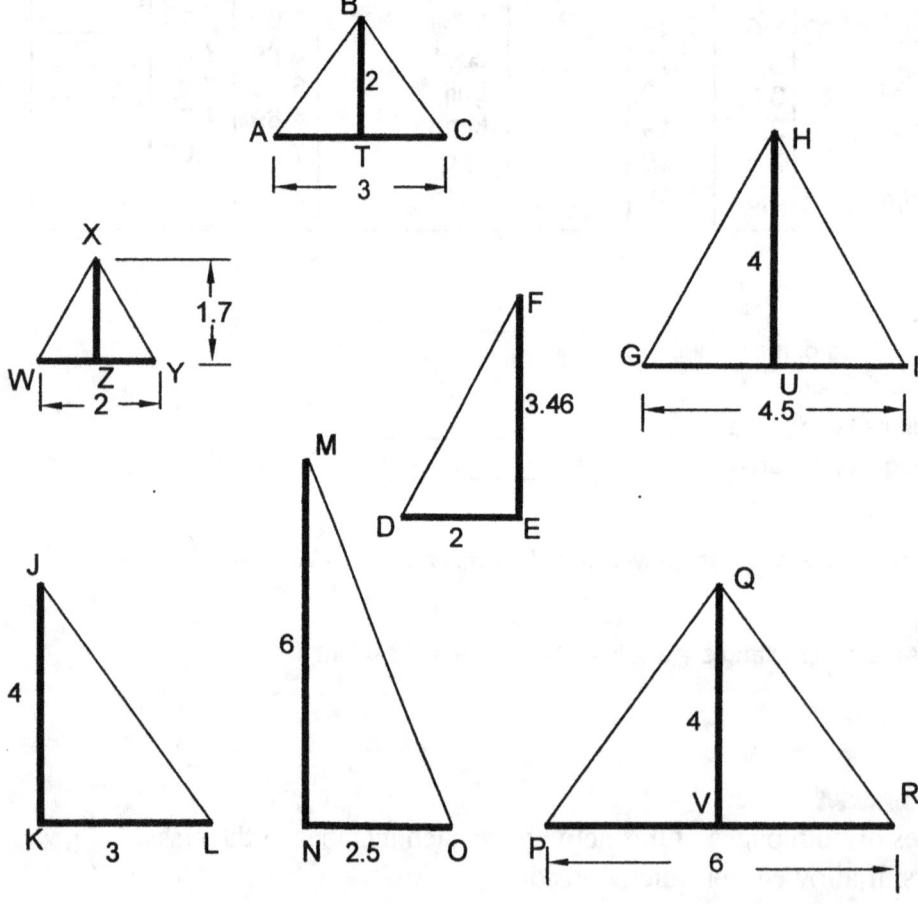

Recall that a polygon is a plane figure that has many (poly) sides (gons). A hexagon has 6 sides, an octagon has 8 sides.

What is a polygon of 9 sides called?

You will find the correct answer to the above question by solving for the area of each of the triangles below. Refer to the figures in the preceding page.

TRIANGLE	AREA	TRIANGLE	AREA
ABC	_____		
WXY	_____	JKL	_____
DEF	_____	MNO	_____
GHI	_____	PQR	_____

What to do

On the line under each circle below, write the area of the triangle indicated above the circle. Example: The area of triangle DEF is 14. Write 14 on the line. Inside the circle, write the 14th letter, N, of the alphabet. Similarly, write the letters of the alphabet corresponding to the areas of the other triangles. What word do the letters in the circles spell out?

DEF	JKL	DEF	WXY	GHI	JKL	DEF
(N)	( )	( )	( )	( )	( )	( )
14	___	___	___	___	___	___

Optional enrichment exercises

ON YOUR OWN
TRY THESE SKILLBUILDERS AND RECALL ENHANCERS

Find the area of:

1. A right triangle whose legs (perpendicular sides) are 12" and 16" long.
2. A right triangle with legs 5" and 12" long.
3. An isosceles triangle with a base 15" long and a height of 20".

Learning goal

To find the area of a triangle when just the lengths of the three sides are given.

Teaching suggestions

* Explain to the class that the square root of a given number is that number which if multiplied by itself yields the given number.
* Calculators may be used, if allowed, to compute the square and square roots of numbers.
* Show as an example how to find the square root of 144.
  Press 144, press 2nd, then press $\sqrt{x}$.
  The answer 12 will show up on the screen.

* Example 2. Find the square root of 100.
  Press 100, press 2nd, then press $\sqrt{x}$.
  The screen shows the answer 10.

Name the formula for finding the area of a triangle when the lengths of its three sides are given.

Directions

The solutions to the following problems will spell out the answer to the above puzzle.

Example: Given, the three sides of triangle ABC. Find its area using only the three sides a, b, and c in the formula:

$A = \sqrt{s\ (s-a)\ (s-b)\ (s-c)}$ where $s = 1/2\ (a+b+c)$

$ = \sqrt{6\ (6-4)\ (6-3)\ (6-5)}$ $= 1/2\ (4+3+5)$

$ = \sqrt{6 \times 2 \times 3 \times 1}$ $= 6$

$ = \sqrt{36}$

$ = 6$ sq. ft.

Triangle diagram: vertex B at top right, vertex A at lower left, vertex C at lower right. Side c - 5', side a = 4', side b = 2'.

Write (spell out) the answer 6 in the boxes at the right. | S | | I | | X |

Spell out the letters in the answers to the problems below:
(Refer to the example on the preceding page.)

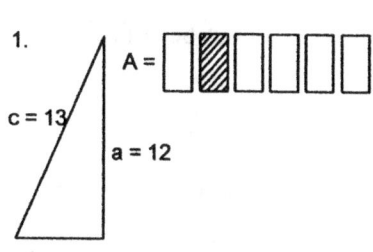

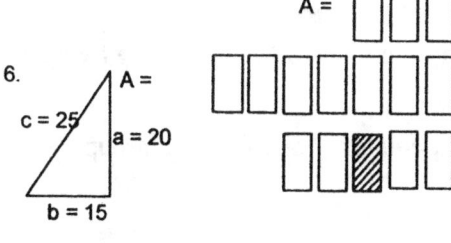

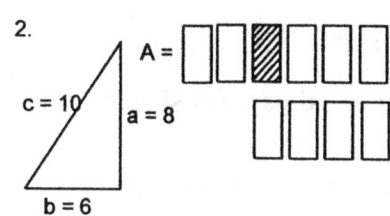

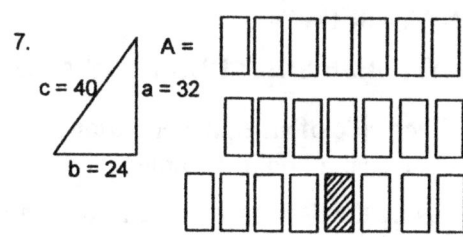

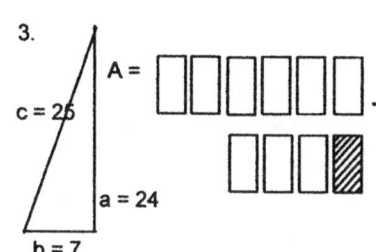

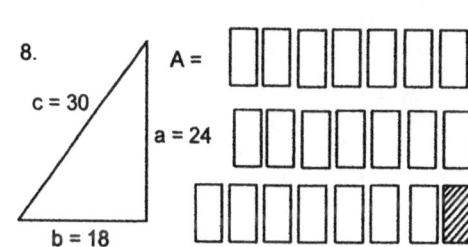

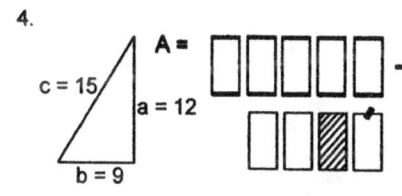

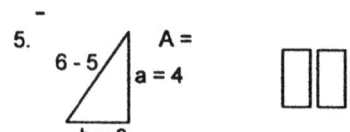

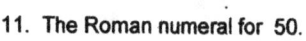

9. The Roman numeral for 1000.

10. The 21st letter of the alphabet.

11. The Roman numeral for 50.

12. The first letter of the alphabet.

Copy on the blanks below the letters in the shaded boxes above from Nos 1 to 12.

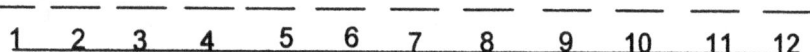

This is the name of the formula to use when only the three sides of a triangle are given.

Prepare ahead

Photocopies of the preceding two pages of this activity for each member of the class

Calculators, one for each group

Optional enrichment exercises

ON YOUR OWN

TRY THESE SKILLBUILDERS AND RECALL ENHANCERS

1. Each side of an equilateral triangle is 10 inches long. Find its area using the 3-sides method just learned.

2. Using the 3-sides method, find the area of a triangle whose sides have lengths of 10", 14", and 20".

Learning goal

Students will be able to find the length of the longest side (hypotenuse) of a right triangle.

Suggestions for teaching

* Explain to the class that the longest side of a triangle is always opposite the right angle. The hypotenuse is often designated by the letter c.

* The perpendicular sides of the right triangle are called legs. The legs are often designated as b and a.

* If the lengths of the legs are given, how could the length of the hypotenuse be determined?

* Recall to the class the Pythagorean Theorem which says that $a^2 + b^2 = c^2$. Show examples. Calculators may be allowed for determining squares and square roots of numbers.

Example 1. In right triangle STU, leg ST = 6"
and leg TU = 8". How long is the
hypotenuse SU?

Solution Given, b = 6" and a = 8". We apply the

Pythagorean Theorem to find c.

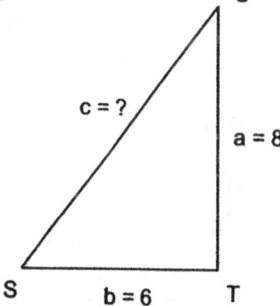

$$a^2 + b^2 = c^2$$
$$8^2 + 6^2 = c^2$$
(Note: It must be explained that
8^2 means 8 x 8 = 64
and 6^2 means 6 x 6 = 36)

$$36 + 64 = c^2$$

$$100 = c^2$$

(What number multiplied by itself equals 100. The answer

is 10, hence c = 10").

On the calculator, press 100, press 2nd, then press x ^{2}.

the result comes out as 10.

Example 2. The legs PQ and QR in triangle PQR are 10" and
12" long, respectively. Find the length of PR.

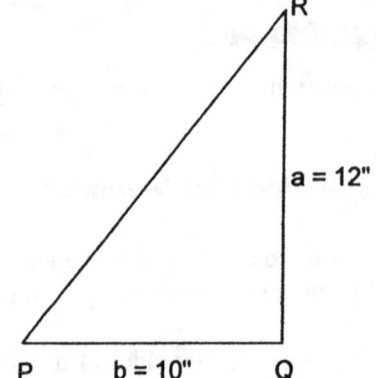

Solution: $a^2 + b^2 = c^2$

$10^2 + 12^2 = c^2$

$100 + 144 = c^2$

$244 = c^2$

$c = 15.620$

On the calculator, press 244, press
2nd, and then press x^2. The answer
comes out as 15.620.

Rounded to 2 decimal places,
the answer comes out as 15.62.

Directions

* Divide the class into groups of 4.

* Every group member rolls the numbers cubes (dice) once for every round of 7 rounds.
  For every roll, the two numbers displayed by the dice will be recorded as values of a
  and   b and recorded in the table below, the higher value should be entered in the a
  column, and the lower valued in the b column.

* Solve for c and c and record their values in the table below.

Roll	a	a^2	b	b^2	c^2	c
1						
2						
3						
4						
5						
6						
7						

Use the calculator to find c in the above table. Remember to round values of c to 2 decimal places.

Prepare ahead

Number cubes (dice) for each group
Photocopies of this activity for each member of the class.

To find the mystery numbers

Add up all the 7 values of c obtained and entered in the last column of the table.
Divide the obtained sum by 7.. Round the result to 2 decimal places.

These 2 digits after the decimal point are the mystery numbers of each member of a group. The group member with the HIGHEST SUM of the two mystery numbers is the winner of the group.

The group having the HIGHEST SUM of the two mystery numbers is the class winner.

Objective

To better understand the properties of two special right triangles - the 30 - 60 - 90 and the 45- 45 - 80 triangles.

Teaching tips

Review the following features and properties of these two special right triangles.

a. The 30-60-90 right triangle:
The longer leg, which is always opposite the 60 degree angle, is always 1.7 times as long as the shorter leg.

In the figure, b = 1.7 times a

The longest side (hypotenuse) is always twice as long as the shorter leg.
In the figure, c = 2 times a.

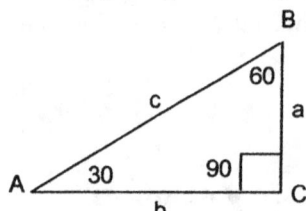

b. The 45- 45- 90 right triangle:
The two legs have equal lengths
In the figure, a = b.

The longest side is 1.4 times as long as the other leg.
c = 1.4 times a, or c = 1.4 times b

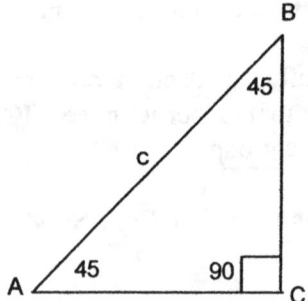

Examples

1. In a 30-69 degree triangle, the hypotenuse is 20 inches long. Find the length of the two other sides (legs).

Solution:
We know that c = 2a, thus 20 = 2a, hence, a = 10.
Also, since b = 1.7a, b = 1.7 times 10 = 17.

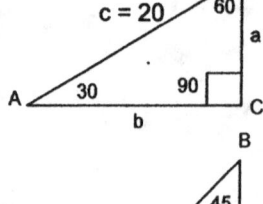

2. In a 45-45 degree triangle, leg a = 10 inches long. Determine the length of the hypotenuse and the other leg.

Solution:

Since c = 1.4a, then c = 1.4 x 10 = 14; also, a = b, hence b = 10.

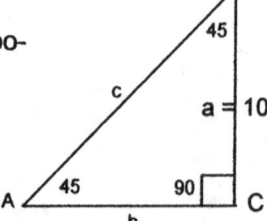

Crossnumber Puzzle

This puzzle is really an application of the properties of the special right triangles.

Teaching tips

Review the class on the following features and properties of the special right triangles.

Across

Given, triangle ABC

A. if c = 675,
   a = _____

C. if a = 1680,
   b = _____

F. if c = 980,
   b = _____

G. if a = 150,
   b = _____

Given, triangle DEF

M. if DF = 804,
   DE = _____

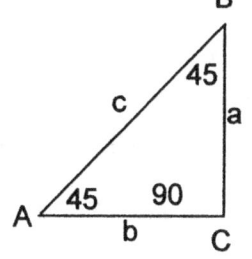

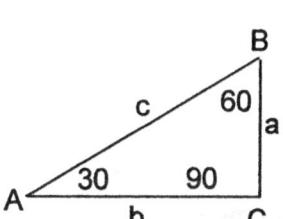

Down

Given, triangle ABC

A. if c = 875,
   a = _____
C. if c = 1680,
   b = _____
F. if a = 980,
   b = _____
H. if a = 150,
   c = _____

Given, triangle DEF

L. if DE = 607
   DF = _____
P. if EF = 520
   DF = _____

 **-12**

No other triangles like these

Optional enrichment exercises

ON YOUR OWN
TRY THESE SKILLBUILDERS AND RECALL ENHANCERS

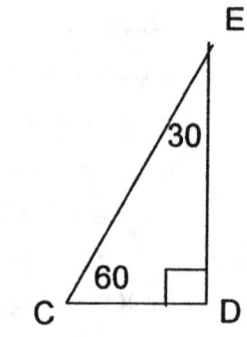

1. Given, triangle CDE at the right. If DE

    = 20 ft,   CD = _____ ft, and CE

    = _____ ft.

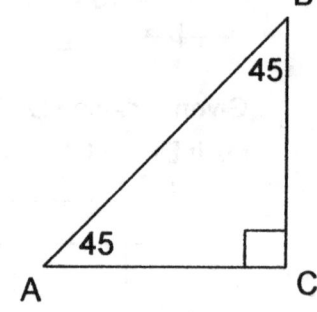

2. AC = 60 ft in triangle ABC at the right.

    AB = _____ ft, and BC = _____ft.

 **13**

Learning goal

The class will gain more skill in identifying quadrilaterals and solving for their areas.

Suggestions for teaching

* Review the class on the various shapes of quadrilaterals namely, the rectangle, square, parallelogram, rhombus and trapezoid.

A rectangle has four sides, two pairs both parallel and equal, and 4 right angles.

A square has 4 sides equal, and has 4 right angles.

A rhombus (diamond-shaped figure) has 4 equal sides, and its angles need not be right angles.

A parallelogram has two pairs of sides both parallel and equal, and its angles need not be right angles.

A trapezoid has four sides - two parallel (called bases) and the other two sides are not parallel.

* Review the area formulas for these quadrilaterals as shown below:

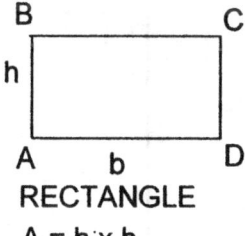

RECTANGLE
A = b x h

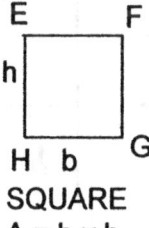

SQUARE
A = b x b

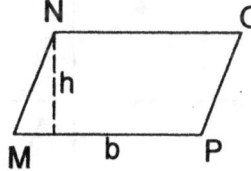

PARALLELOGRAM
Area = b x h

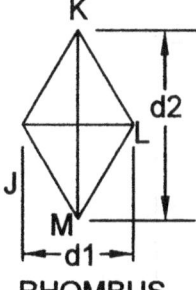

RHOMBUS
Area = d1 x d2

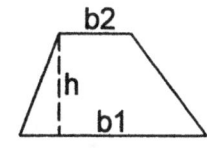

TRAPEZOID
A = 1/2 (b1 + b2) x h

Problem

Find the area of the following quadrilaterals. Write their areas in the square at the bottom of this page, and arrange them in such a way that when added VERTICALLY, HORIZONTALLY, or DIAGONALLY, the sum is always 18.

	b	h	Area
Rectangle A	3	2	___
Rectangle B	4	2	___
Square C	2	2	___
Square I	1	1	___
Square E	3	3	___
Parallelogram F	7	1	___
Parallelogram G	5	1	___

Trapezoid H

Rhombus D   d1 = 2   d2 = 3

Area
___
___

b1 = 8   b2 = 3   h = 2

D	F	B
H	A	I
C	G	E

 **13**

What to prepare ahead

Photocopy of page 2 of this activity for each member of the class.

Optional enrichment exercises

ON YOUR OWN
TRY THESE SKILLBUILDERS AND RECALL ENHANCERS

Solve for the areas of the following quadrilaterals:

AREA

1. Square, one side s = 12. _____

2. Rectangle, b = 10, h = 8. _____

3. Rhombus, d1 = 6, d2 = 8 _____

4. Trapezoid, b1 = 10, b2 = 6, h = 8 _____

5. Parallelogram, b = 12, h = 10. _____

(Note: All the above dimensions are in inches.)

 **14**

Objective
To construct a bridge model designed along the shapes of triangles and rectangles.

Directions
Students will build a bridge model using soda straws and scotch tape.

Divide the class into groups of 3: one group member to measure desired lengths on the straws, a second one to cut the straws, and a third member to tape the straws together into the desired structure patterns;

When finished, the bridge models will be positioned with both ends resting on books (see the figure on page 3 of this activity);

Place on top of the bridge an index card, then place 5 washers on the index card.
Keep adding one more washer after another until the bridge will give way or collapse.

Place an index card on top of the bridge, then place 5 washers on the card;

Keep adding one more washer after another until the bridge gives way or collapses;
The bridge carrying the most number of washers, at the point of collapse of the bridge, WINS

Materials used
Soda straws, scotch tape, scissors, ruler or measuring tape.

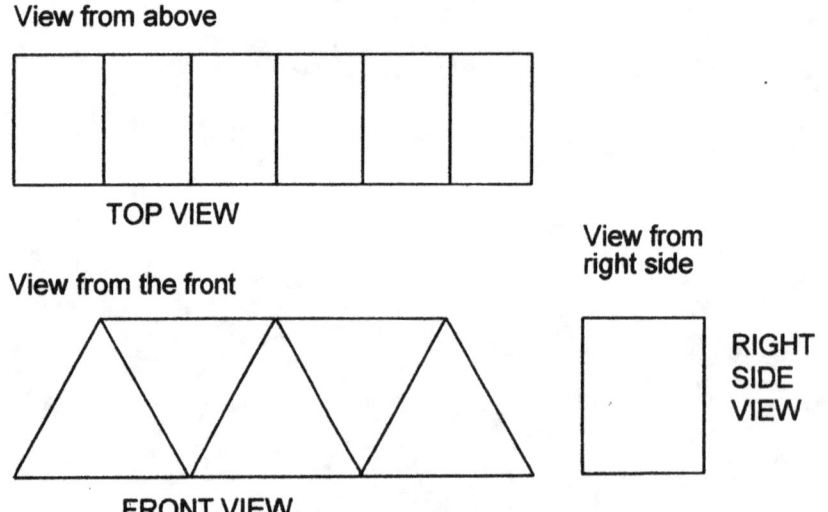

View from above

TOP VIEW

View from the front

FRONT VIEW

View from
right side

RIGHT
SIDE
VIEW

Each side of triangle ABC has a length of 2.5 inches.

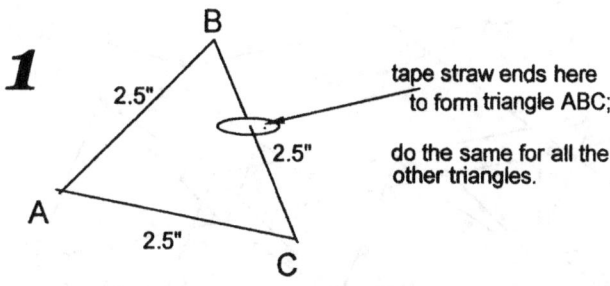

1

B

2.5"

2.5"

A

2.5"

C

tape straw ends here
to form triangle ABC;

do the same for all the
other triangles.

2

tape together triangles
ABC and BCD

B B

D

A

C C

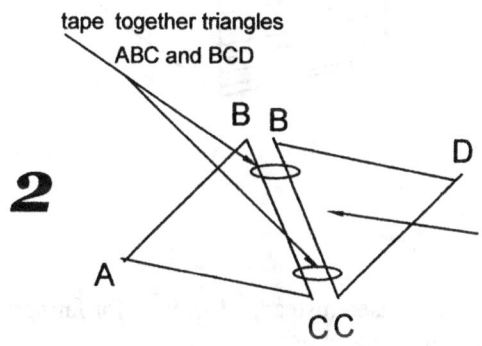

3 Assemble 5 triangles to form one side of the bridge
and the other 5 triangles to form the other side.

Attach the 7 cross members
to the triangles as shown at
the typical joint at the right.
Use tape.

4

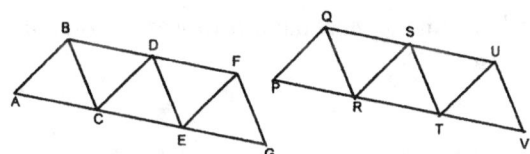

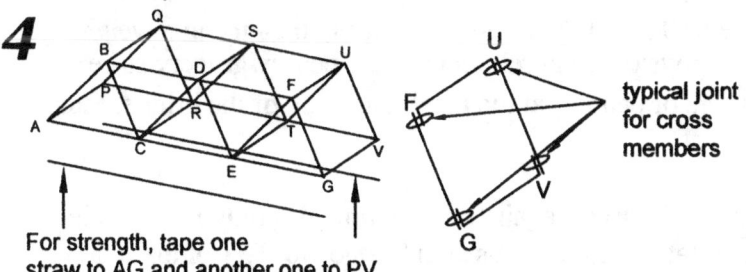

typical joint
for cross
members

For strength, tape one
straw to AG and another one to PV

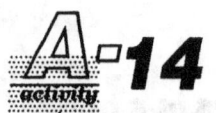
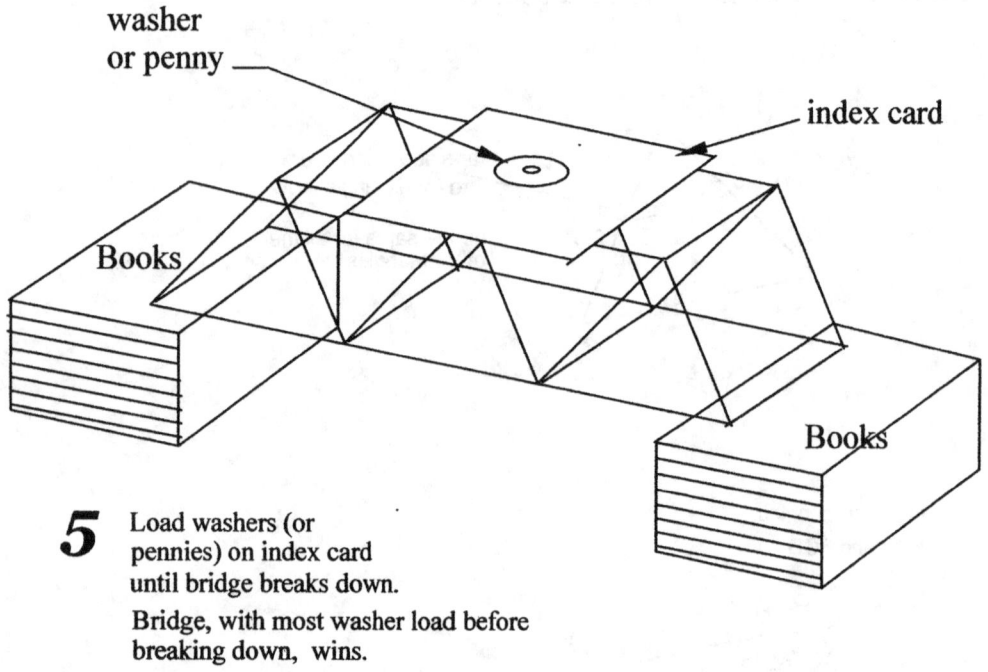

washer
or penny

index card

Books

Books

5 Load washers (or
pennies) on index card
until bridge breaks down.

Bridge, with most washer load before
breaking down, wins.

NOTE:

If washers are not available, coins such as PENNIES may be used instead. However, for fairness
and uniformity, the class should decide which one to use for all groups.

The bridge with the most number of washers on it at the breaking point of the bridge (or the point
of collapse) wins.

ON YOUR OWN
EXPLORE GREATER POSSIBILITIES

Build your own bridge using soda straws.

As in this activity, load the bridge with washers or pennies, then using a weighing
scale from the chemistry or physics lab at school or at the nearby grocery store.
find the weight of the washers or pennies at the breaking point of the bridge. Make
a record of this weight.

Build a second bridge using soda straws again, but this time use right triangles in
the design instead of the equilateral triangles used in bridge No. 1. Compare the
breaking-point loads in both cases and make a conclusion. Which one of the two
bridges is the stronger structure?

 **15**

Learning goal

Better understanding of the meaning of pi, its origin, and its practical value in real life.

Directions

This activity is an experiment to determine the value of pi. (The mathematical symbol of pi is the Greek letter π).

* Divide the class into groups of 4.
* Have each group measure the circumference C of a penny using a tape meassure. Another way to measure the circumference of the penny is to make it roll on paper then measuring the distance traveled in one turn. This is the circumference of the penny.
* Measure the diameter D of the penny as well.
* Measure also the C and D of the nickel, dime, quarter and the other object (s).
* Find the C / D values for each of the coins and objects.
* Record all the readings in the table below.

* For the unnamed object in line 5 of the table below, each group selects one from the following:

half-dollar coin	soft drinks bottle	coffee mug
dollar coin	any round bottle	drinking glass
soft drinks can	small saucer	flashlight battery

Record also the C, D, and C / D values in the table.

Object	Circumference C	Diameter D	C / D
1. Penny	_____	_____	_____
2. Nickel	_____	_____	_____
3. Dime	_____	_____	_____
4. Quarter	_____	_____	_____
5. _____	_____	_____	_____

Record the C/D values in the last column of the table.

Evaluating the data

(Note: Calculators, if allowed, may be used for the following divisions. Round the answers to 4 decimal places).

* For each of the objects, divide C by D to obtain the C/D values.

* Add all the C/D values. Divide the sum by 5 to get the average C/D value;

* The group with the average C/D value closest to the accepted value of pi which is 3.1416 wins the game.

What to prepare ahead

Photocopy of page 1 of this activity for each group
Tape measure / ruler for measuring the circumference and diameter of the objects.

Optional enrichment exercises

ON YOUR OWN

TRY THESE SKILLBUILDERS AND RECALL ENHANCERS

Given, C / D = 22/7.

Find C, if:

1. D = 20"   2. D = 12"   3. D = 15"

Find D, if

4. C = 20π "   5. C = 40π "   6. C = 60π"

Objective

The class will learn more about the circumference and area of circles.

Directions

* Hand each student a photocopy of the next two pages of this activity.

* Ask each student to paste the picture on the next page on either poster board or manila folder.

* Draw 9 equally-spaced vertical lines and 4 equally-spaced horizontal lines on the picture.

* Cut along the vertical and horizontal lines to produce 36 rectangles which are really 36 picture blocks of a 36-piece picture puzzle.

* Solve all the problems on page 3 of this activity.

* Arrange all the 36 picture blocks over the solution guide to form the original picture. Write on the back of each block the answer to the problem covered by the block.

* Proceed similarly with the other 35 blocks.

    * Circumference C of a circle: $C = \pi \times D$
      Diameter D of a circle   : $D = C / \pi$

    * Area A of a circle   : $A = \pi \times r^2$

Example 1.   Given, the circumference of a circle is $10\,\pi$. Find the diameter.
          Solution: $C = \pi \times D$
                    $10\,\pi = \pi \times D$, hence $D = 10$.

Example 2.   Given: the diameter D of a circle is 6. Find the circumference C.
          Solution: $C = \pi \times D$, hence $C = 6\pi$

Example 3:   Given: the radius r of a circle is 10. Find its area A.
          Solution: $A = \pi \times r^2$, hence $A = \pi \times 100 = 100\pi$

Example 4.   Given: the area A of a circle = $64\,\pi$. Find its radius r.
          Solution: $A = \pi \times r^2$
                 $64\pi = \pi \times r^2$, $64 = r^2$, hence $r = 8$.

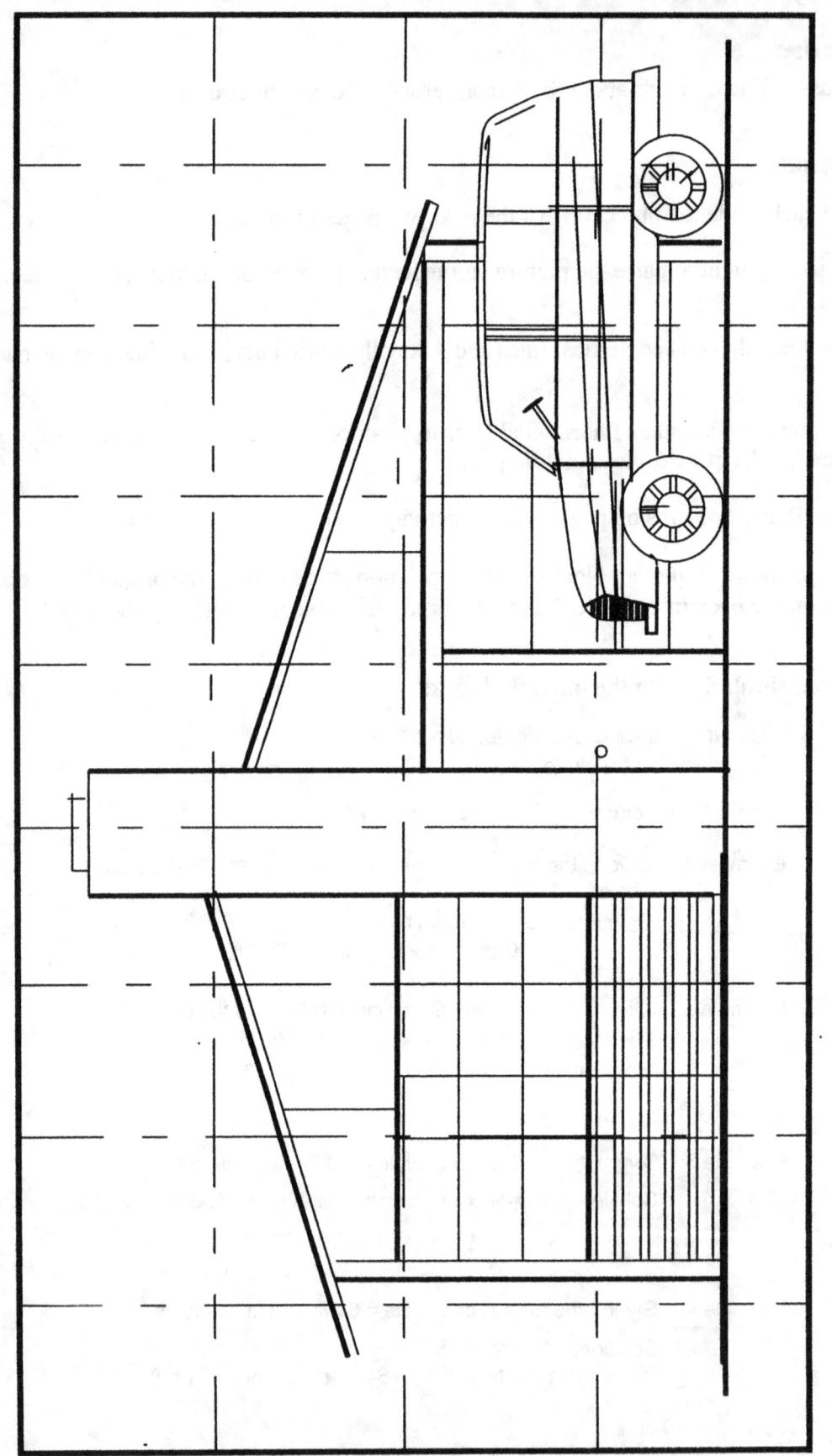

B

$C = 4\pi$	$D = 4$	$A = 100\pi$	$A = 16\pi$	$C = 16\pi$	$D = 22$	$A = 49\pi$	$D = 20$	$C = 24\pi$
$D =$	$A =$	$r =$	$r =$	$D =$	$A =$	$r =$	$C =$	$D =$
$D = 6$	$A = 4\pi$	$C = 6\pi$	$D = 6$	$D = 4$	$D = 30$	$A = 64\pi$	$C = 18\pi$	$D = 18$
$C =$	$r =$	$D =$	$A =$	$C =$	$A =$	$r =$	$D =$	$C =$
$D = 2$	$A = 9\pi$	$D = 8$	$A = 25\pi$	$A = 36\pi$	$C = 16\pi$	$A = 8\pi$	$D = 20$	$D = 24$
$A =$	$r =$	$C =$	$r =$	$r =$	$D =$	$r =$	$A =$	$A =$
$C = 10\pi$	$D = 12$	$D = 8$	$C = 12\pi$	$D = 14$	$D = 14$	$D = 16$	$C = 20\pi$	$D = 24$
$D =$	$C =$	$A =$	$D =$	$A =$	$A =$	$C =$	$D =$	$C =$

A

Conclusion

The outcome of this activity is two-fold; (1) it has produced a picture puzzle, and (2) it is also a skillbuilder device in the sense that it provides opportunity for drill and mastery of 36 problems relating to circles.

The solution guide base gives the problem and the answer to that problem is right there at the back of the picture block.

Learning goal

Investigate how and if triangles could form circles.

Suggestions for teaching

Review the class on the area A of a triangle, and that A = 1/2 x base x height.

With figures and illustrations, show that infinitely many triangles that are placed next to each other side by side, with their vertices placed together at a common point, would ultimately form a circle.

1

One triangle with base b and height h inside the circle.

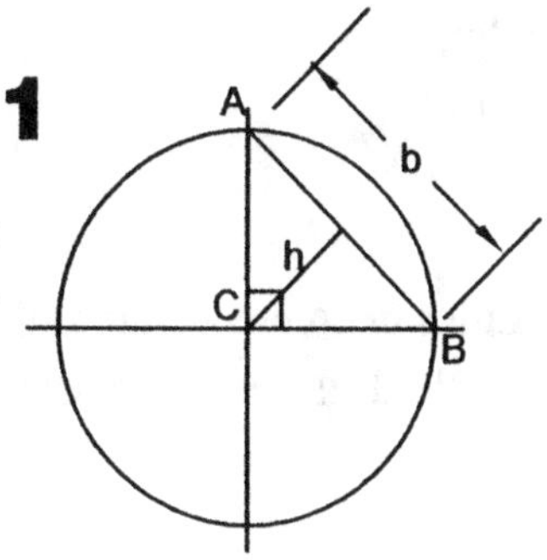

2

Four triangles inside the circle

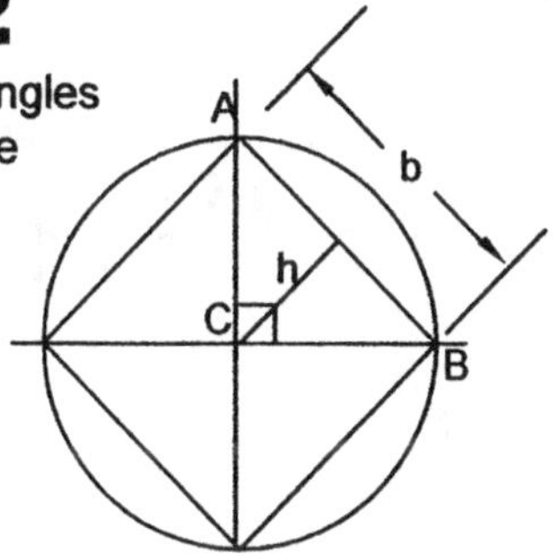

3

Eight triangles inside the circle

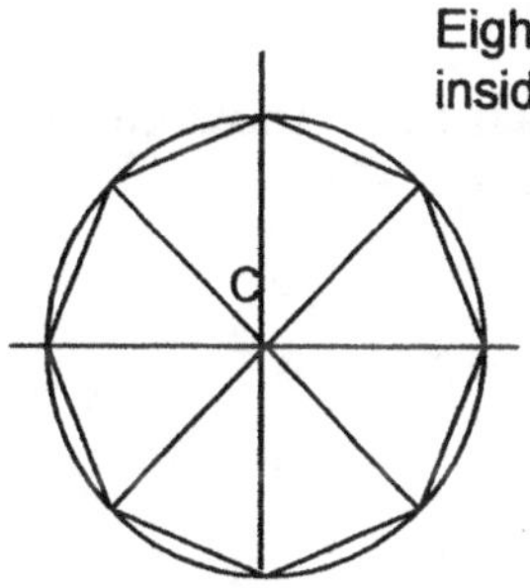

4

Sixteen triangles inside the circle

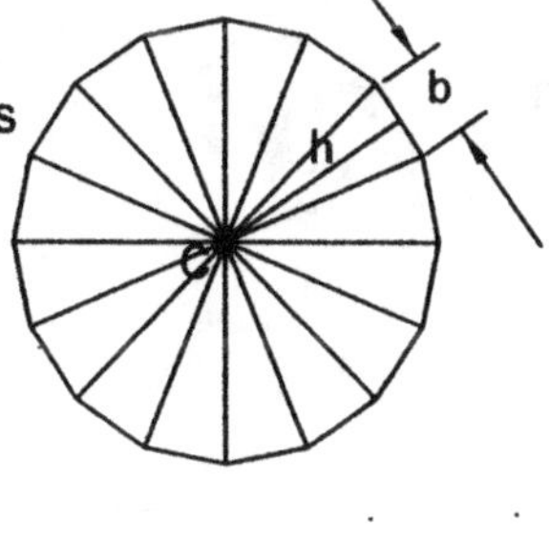

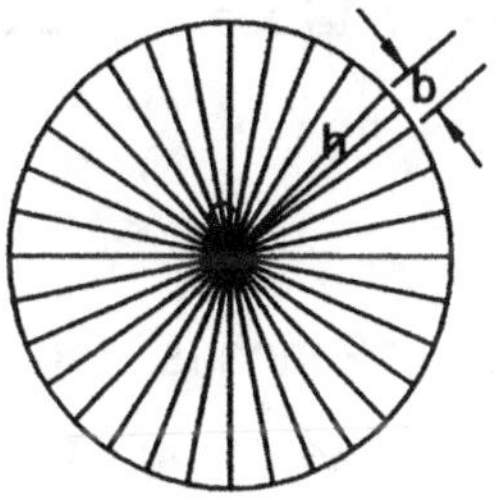

5

Thirty-two triangles inside the circle d

The big question | What happens if the number of triangles inside the circle is increased to 100 or more?

* The area A of the triangle in Fig. 1 is A = 1/2 x b x h.

* The area A of the triangles in Fig 5 is A = 32 x 1/2 x b x h.

Observe that, as there are more and more triangles, their combined area comes close to the area of the entire circle.

At that point, the height of the triangles becomes equal to the radius of the circle, and the many bases become just little dots that form the circumference C = 2 x pi x r.

Thus, the area of the circle = the area of the several triangles = 1/2 x 2pi x r x r = pi x r square. (A = pi x r²).

Example 1

Find the area of a circle with a diameter of 10 inches.
Solution: The radius = 10 / 2 = 5.
Hence, the area A = pi x r² = 3.14 x 5² = 78.5 sq. in.

Example 2

Given, a circle with a radius of 12 inches. Find its area.
Solution: A = pi x r² = 3.14 x 144 = 452.16 sq. in.

The crossnumber puzzle on the next page presents exercises on the area

of circles for building skills and enhancing recall power.

A		K		B		
		P			N	
	M					
R			S		U	
	T					
						V

(All the following problems deal with the circle. Take 3.14 as the value of pi. All answers must be rounded to the nearest one or nearest unit. Example: if the answer turns out to be 473.94, it should be rounded to 474.)

ACROSS

A. d = 20, area = _____

B. r = 12, area = _____

P. r = 10, dia. = _____

N. r = 16, dia. = _____

K. dia. = 48, r = _____

U. dia. = 42, r = _____

T. dia. = 30,

    area - 376 = _____

DOWN

A. dia. = 20, r = _____

K. r = 12, area = _____

B. r = 21, dia = _____

N. r = 161, dia = _____

M. r = 12, dia. = _____

R. 200 pi - 400 = _____

S. 80 pi - 12 = _____

V. r = 10, area - 214 = _____

 **18**

Learning goal

To learn more about the properties of the circle.

Teaching notes

Review the class on the meaning of pi (π), that the circumference C of a circle divided by the diameter d , or that $C/D = \pi$.

* Thus, if $C/D = \pi$, then $C = \pi \times d$.

* If A is the area of a cicle, $A = \pi \times r^2$.

* The radius r is always equal to 1/2 of the diameter d or $r = 1/2 \times d$.

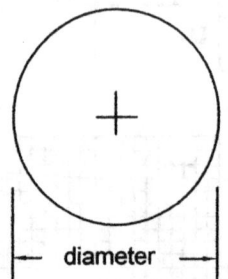 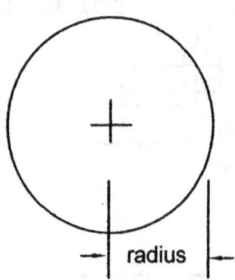

diameter   radius

Example 1

Given, a circle with a diameter of 12 inches. Solve for its circumference, leave your answer in terms of π , and round it up to 2 decimal places.

Solution: $C = \pi \times D$
$= \pi \times 12$
$= 12\,\pi$ in.

Example 2

A circle has a diameter of 2 inches. Find its area in terms of π , and round the answer to 2 decimal places.

Solution: $A = \pi \times r^2$
$= \pi \times 10^2$
$= 100\,\pi$ sq. in.

Example 3

Find the circumference and area of a circle with a diameter of 15 ft. Round your answer to 2 decimal places, and leave them in terms of π

Solution: $C = \pi \times d$ $A = \pi \times r^2$
$= \pi \times 15$ $= \pi \times 7.5^2$
$= 15\,\pi$ ft $= 56.25\,\pi$ sq. ft.

Directions

* Divide the class into groups of 4.

* With the Target Board (shown below) taped to the floor, group members take turns in dropping a cone (shown in the next paragraph below) onto the Board.

* Where the cone lands on the Board, mark a point P1 directly below the cone's vertex V. Obviously, P1 is where the cone's vertex hit the Board. The points in the next 4 drops (each member makes 5 drops) should be marked P2, P3, P4, and P5.

cone ───── Mark this point P1 on the target board
V

> Note: If P1 lies on a strip between two adjacent circles, find the area of each of the two circles, then subtract the lesser from the bigger area. This is the area of the strip. Record this area in the table on the next page.

* If the cone lands on a circle and V lies on that circle, write 10π in the table.
* If V lies inside the smallest circle (bull's eye), write 5π in the table.
* The player with the LOWEST TOTAL SCORE is the WINNER. (Why lowest? Explain). Yes, lowest, because the lowest score means that a player has been hitting closest to the bull's eye in the center.)

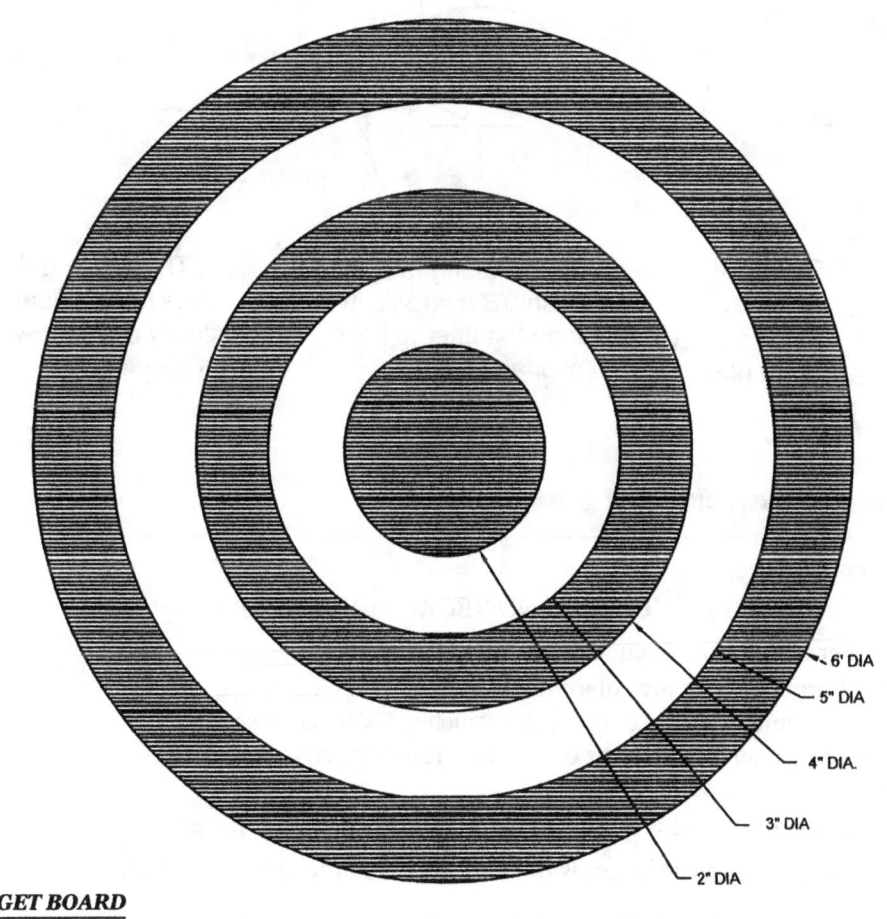

6' DIA
5" DIA
4" DIA
3" DIA
2" DIA

TARGET BOARD

Area Table

DROP	BIG CIRCLE AREA	SMALL CIRCLE AREA	CIRCLE	STRIP AREA
1				
2				
3				
4				
5				
TOTAL STRIP AREA ⟶				

Prepare ahead

Photocopy of the Target Board for each group
Drop cone for each group (figures below show how to construct a drop cone)

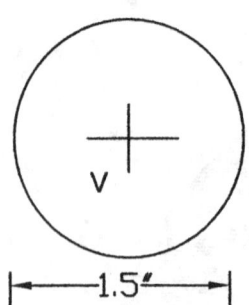

|←——1.5″——→|

1. Cut a circle 1.5" in diameter from plain white paper.

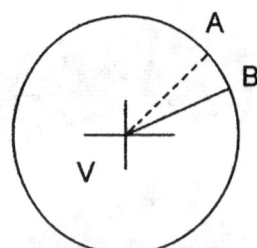

2. Cut along radius VA, push VB over VA, then tape together to hold overlap in place.

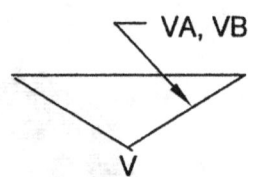

3. The finished drop cone should float slowly on air down to target board.

Optional enrichment exercises

ON YOUR OWN
TRY THESE SKILLBUILDERS AND RECALL ENHANCERS

1. The formula for the circumference of a circle is _____

2. The formula for the area of a circle is _____

3. The circumference of a circle is 20 π inches. What is its area?
   (Leave your answer in terms of π, also round the answer to 2 decimal places). Answer _____

4. If the area of a circle is 20 square inches, what is its circumference?
   (Leave your answer in terms of π and round the answer to 2 decimal places). Answer _____

19

Activity goal

To find the volume of a cone.

Suggestions for teachers

* Review the class on the parts of a cone, namely: base, radius, height or altitude.

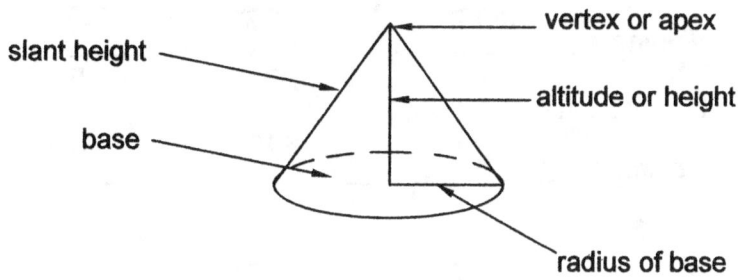

slant height

base

vertex or apex

altitude or height

radius of base

* Show examples of how to find the volume V of a cone.

* Show examples of how to find the volume V of a cone.

Example 1. The diameter D of the base of a cone is 10 inches. If its height h is 6 inches, what is its volume V?

Solution: $V = 1/3 \times \pi \times r^2 \times h.$
$= 1/3 \times \pi \times 5^2 \times 6.$
$= 50 \pi$ cu. in.

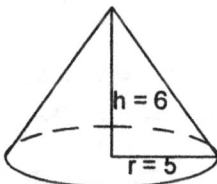

h = 6

r = 5

Example 2. Given, a cone with an altitude of 10 inches. If the radius of the base is 3 inches, what is the volume of the cone?

Solution: $V = 1/3 \times \pi \times r^2 \times h.$
$= 1/3 \times \pi \times 3^2 \times 10$
$= 30 \pi$ cu. in.

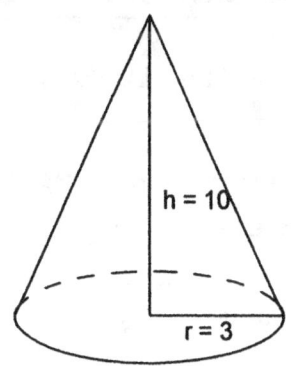

h = 10

r = 3

Name the magician who wears a cone-shaped hat.

The answer to the above question may be found by solving the volumes
of the cones with the given dimensions below:

D = 20", h = 9"

Volume = _____300 π_____ , _____200 π_____ , _____100 π_____ , _____60 π_____ .
             M   N   O   P

D = 12", h = 10"

Volume = _____130 π_____ , _____120 π_____ , _____160 π_____ , _____180 π_____
             D   E   F   G

D = 18". h = 10"

Volume = _____200 π_____ , _____60 π_____ , _____270 π_____ , _____240 π_____ .
             P   Q   R   S

D = 6", h = 5"

Volume = _____30 π_____ , _____20 π_____ , _____15 π_____ , _____12 π_____ .
             J   K   L   M

D = 8", h = 6"

Volume = _____32 π_____ , _____24 π_____ , _____18 π_____ , _____16 π_____ .
             I   J   K   L

D = 8*, h = 12"

Volume = _____54 π_____ , _____64 π_____ , _____44 π_____ , _____84 π_____ .
             M   N   O   P

Write on the blanks below the letters under the correct answers above.
to find the desired response to the question above.

_____ _____ _____ _____ _____ _____

Optional enrichment exercises

ON YOUR OWN
TRY THESE SKILLBUILDERS AND RECALL ENHANCERS

Find the volume of the cones below:

Base diameter = _____6"_____ , height = _____4"_____

Base radius = _____5"_____ , height = _____12"_____

Base diameter = _____12"_____ , height = _____8"_____

Base radius = _____10"_____ , height = _____24"_____

Learning goal

Identify the x and y coodinates of a point on the coordinate plane.
Locating points on the coordinate plane.

Suggestions for teaching

Draw on the board or on the overhead projector the x-axis and the y-axis. Show that these two lines (number lines) lie on a plane called the coordinate plane as shown below,

Example:

1. Locate point P(10,14) on the coordinate plane.

   The number 10, called the x -coordinate, means that 10 units be counted along the x axis; the number 14, called the y-coordinate, means that 14 units be counted along the y-xis.

2. Name the coordinates of point Q.
   There are 20 units counted along the x-axis, and 24 units along the y-axis.
   Hence, the coordinates of point Q are (20, 24).

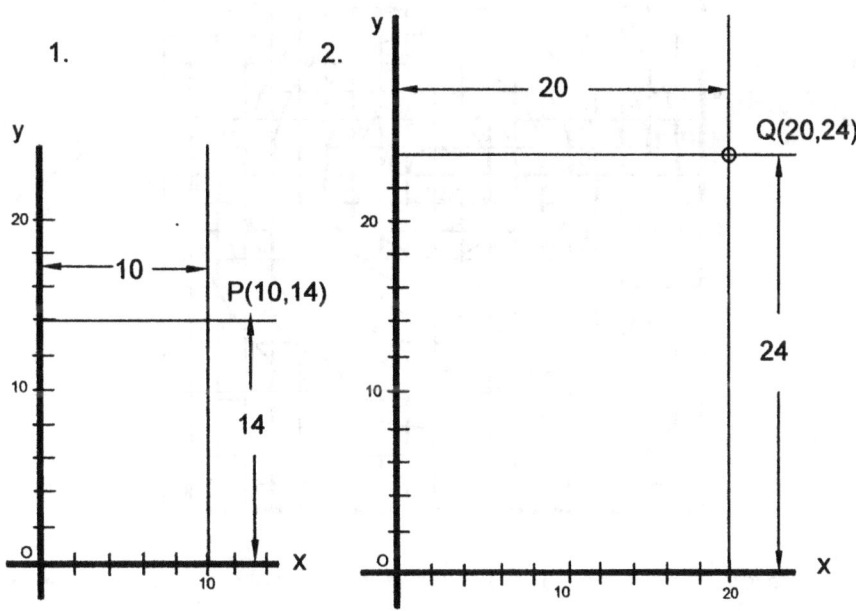

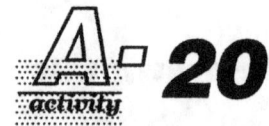

What could Robin Hood have done with his arrows on the target board?

Directions

Robin Hood was the sharpest shooter in Sherwood Forest. He was the wizard of the bow and arrow, the champion archer in all of England in his time. Little John, Friar Tuck, and Will Scarlet were among the many in Robin's band of merry men in Sherwood Forest.

To get the correct response to the question above, find the answers to the questions below:

To get the correct response to the riddle above, find the answers to the questions below.

* Mark the points whose coordinates are as follows:

A (6, 32)	D (20, 16)	G (24, 32)
B (6, 24)	E (24, 16)	H (20, 32)
C (6, 16)	F (24, 24)	I (20. 24)

* Connect A to B, B to C, C to D, D to E, E to F, F to G, G to H, and H to A. Also connect A to E and C to G. Finally connect B to F, and D to H.

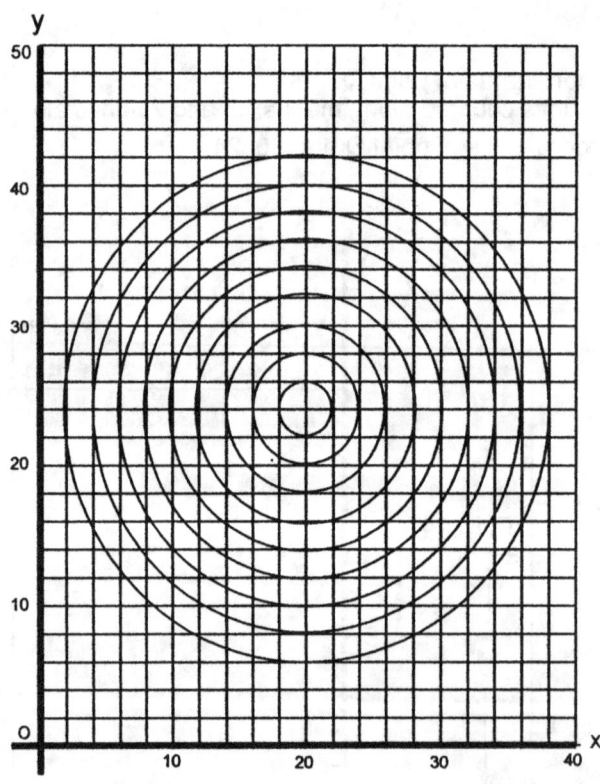

The figure formed by connecting the points as indicated above would turn

out to be what? _____

Prepare ahead

Graphing paper for each member of the class
Graphing paper transparency for use on overhead projector

Optional enrichment exercises

ON YOUR OWN
TRY THESE SKILLBUILDERS AND RECALL ENHANCERS

1. Plot (mark) these points on the coordinate plane. Use graph paper.
   A (2, 2)   B (4, 4)   C (6, 6)   D (8, 8)   E (10, 10)

2. When the above points are connected together, what figures do they form?

3. Locate the following points on the coordinate plane. Use graph paper.

   P (20, 34)   D (20, 26)
   B (12, 30)   E (28, 18)
   C (12, 18)   F (28, 30)

4. Connect together the above points and describe the figure formed.

Learning goal

To construct a cone-shaped hat for a "Mardi Gras" costume.

Directions

* A half-size piece of posterboard, or one manila folder, may be used for this construction project.

* Scissors and tape will be needed for this project

Step 1. On the top of the posterboard, draw a circle with a diameter of 8 inches (this diameter size should reasonably fit most heads in class). Use a compass to draw the circle, or you may use a paper clip, string and pencil if a compass is not available.

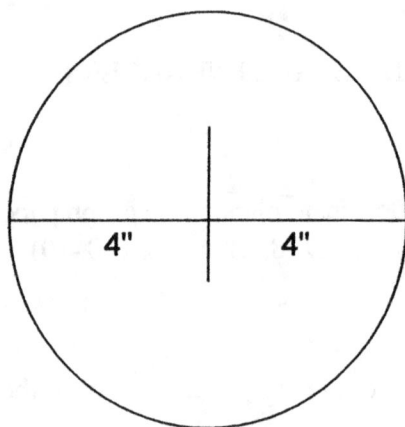

Step 2. Divide the circle into 16 divisions as shown below. Mark the endpoints of two diameters as points A and B.

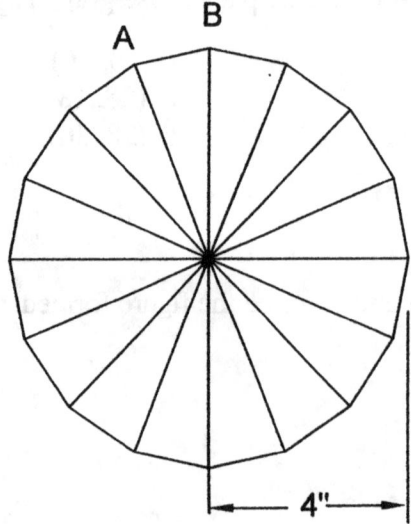

Step 3. The hat being constructed here has the shape of a cone. Estimate
its height to be 8 inches.

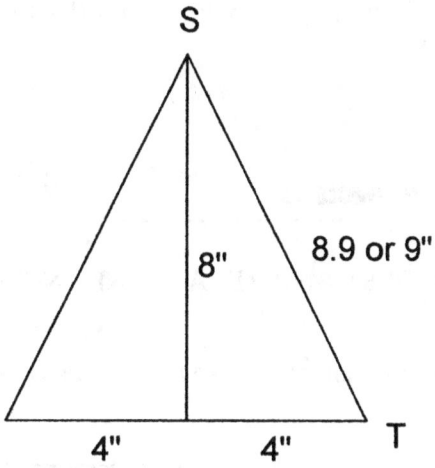

Step 4. Draw another circle with a radius equal to the 9" length of the side ST
in the above figure. (Note: ST is called the slant height of the cone).

Starting at point P, draw 16 equal divisions on the circle shown below, (the length of each division being equal to AB in Step 2). Provide a flap, as shown, for the glue to hold the sides together and form
the cone.

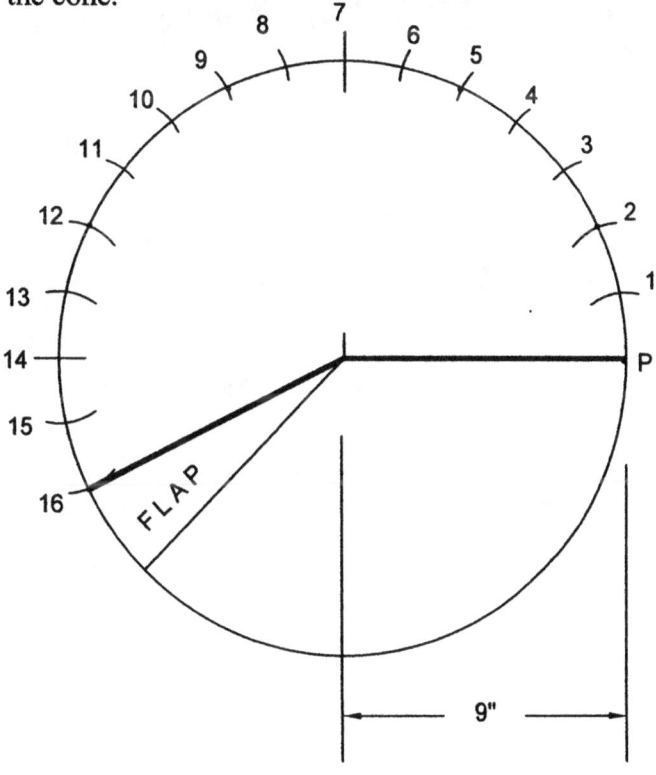

Step 5. Using pencil crayons, pentel pen, ball pen or pencil, decorate the side of the finished hat with your original design or any form of decorative artwork.

The teacher decides which is the best-looking hat - (best construction and best artwork design).

Optional enrichment exercises

ON YOUR OWN
TRY THESE SKILLBUILDERS AND RECALL ENHANCERS

The fromula for the area A (sometimes called lateral area or area of the side of a cone) is equal to:

A = 1/2 x slant height x circumference of the base.

Find the area of the following cones:

1. base radius = 10", slant height = 13". Leave your answer in terms of pi.

2. base diameter = 6", height (not slant height) = 4". (Leave your answer in terms

Activity goals

Students will learn about the various shapes of geometric solids. They will also learn how to measure their volumes and use these calculations and skills for practical purposes in real life.

Teaching suggestions

Draw on the board the rectangular solid (rectangular prism), cube, cylinder, and cone. Explain the formula for the volume of each of these solids as shwon. below:

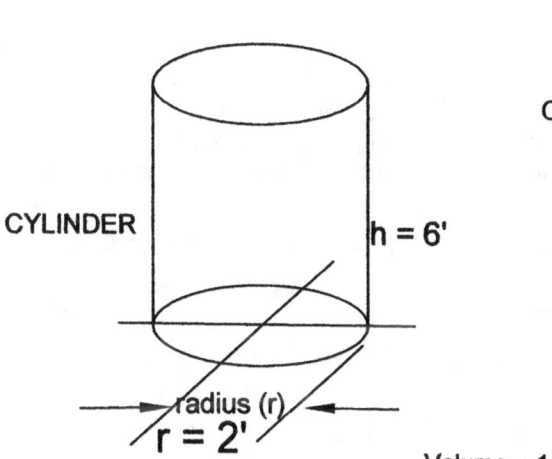

CYLINDER

Volume = area A of base x height
= pi x r^2 x h

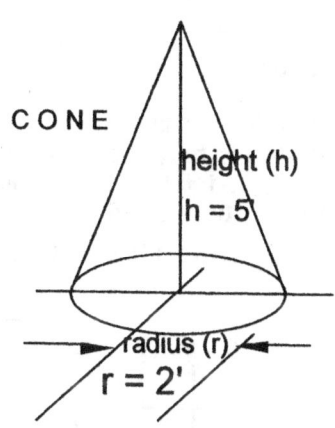

CONE

Volume = 1/3 x area A of base x height
= 1/3 x pi x r2 x h

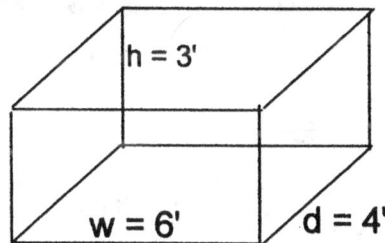

RECTANGULAR PRISM

Volume = w x h x d

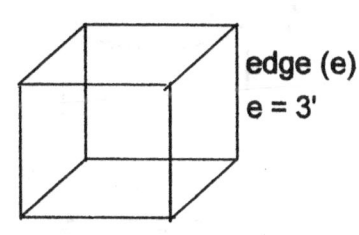

CUBE

Volume = e x e x e or e^3.

How far up above the earth can you fly?

Directions

* Divide the class into teams of 4. Each team pretends to be a space crew flying a spaceship vertically upward as far up as its fuel could bring it

* Members of a team take turns in rolling the number cubes (dice). The team could get a fuel tank if the displayed numbers on the dice add up to: 3, 6, 9, or 12. (Only when the dice display these numbers would a team get a fuel tank)

Referring to figures A, B, C, and D below:
(If a team gets the number 3 from the dice display, the team qualifies for Tank A; if a team gets the number 6 from the dice display, they qualify for Tank B; display number 9 qualifies for tank C; and display number 12 for Tank D). Other numbers do not at all qualify a team for a Tank.

* Each team will have 10 chances to roll the dice. That means they will have, if lucky, a total of 10 tanks for flight. (Note: The flight range is 100 miles for every cubic foot of fuel). The example below will help clarify directions.

Example

Team No. 1 rolls the cubes and the displayed numbers are
termined to be 20.95 cubic feet. This could propel them through
6 and 6, a total of 12. This qualifies the team for Tank D, whose volume is de-
20.95 x 100 = 2095 miles.

* The team traveling the farthest distance up is the class WINNER.

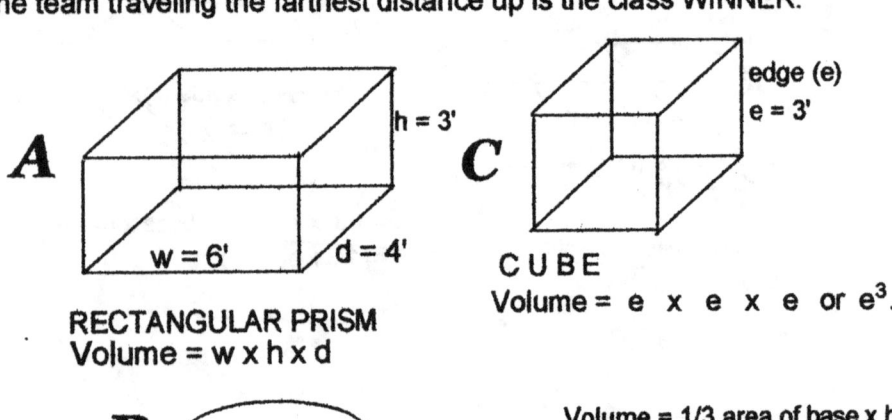

A

h = 3'
w = 6'
d = 4'

RECTANGULAR PRISM
Volume = w x h x d

C

edge (e)
e = 3'

C U B E
Volume = e x e x e or e^3.

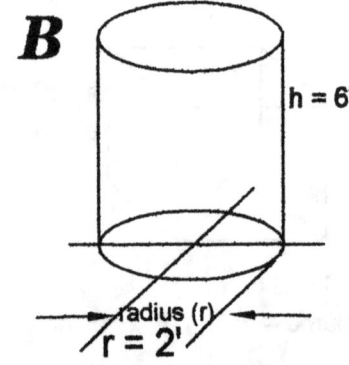

B

h = 6'

radius (r)
r = 2'

CYLINDER
Volume =   area A of base x height
      = pi x r^2 x h

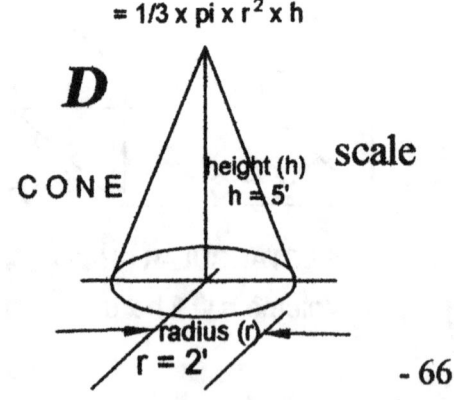

Volume = 1/3 area of base x h
    = 1/3 x pi x r^2 x h

D

C O N E

height (h)
h = 5'

scale

radius (r)
r = 2'

 **22**

Data Table

Volume A = 98 cu. ft.
Volume B = 32 $\widetilde{\pi}$ cu. ft.
Volume C = 27 cu. ft.
Volume D = 6.67 $\widetilde{\pi}$ cu. ft.

NOTE: The data entries in the table below are for the example shown on the preceding page.

ROLL	DISPLAYED NUMBERS		SUM OF DISPLAYED NUMBERS	TANK	VOLUME OF FUEL	TRAVEL DISTANCE
1	6	6	12	D	20.95 cu. ft.	2.095 miles
2						
3						
4						
5						
6						
7						
8						
9						
10						

What to prepare ahead

Photocopies of page 2 of this activity for each member of the class
Numbers cubes (dice), one pair for each group

Optional enrichment exercises

ON YOUR OWN

TRY THESE SKILLBUILDERS AND RECALL ENHANCERS

(Round answers to 2 decimal places and leave them in terms of pi).
Find the volume of
1. A cone, radius of base = 10", h = 18".
2. A cylinder, radius of base = 10", h = 18".
3. A rectangular prism, w = 12", h = 8", d = 10".
4. A cube, e = 20".

Learning Goal

After this activity, the class will better understand the meaning of symmetry and disco-ver countless examples of symmetry around us.

Teaching tips

* Fold a piece of paper, place some ink or water color paint on the fold line, and then press the paper over the ink or paint a number of times to produce a symmetric figure.

* The fold is (a) the line of symmetry, the figure on the left of the fold is called the (b) pre-image, and the figure of the right of the fold is (c) the image. (See figure below.

* The distance of the pre-image points from the line of symmetry is equal to the dis-tance of the image points from the same line of symmetry.

* Some examples of symmetry are shown below:

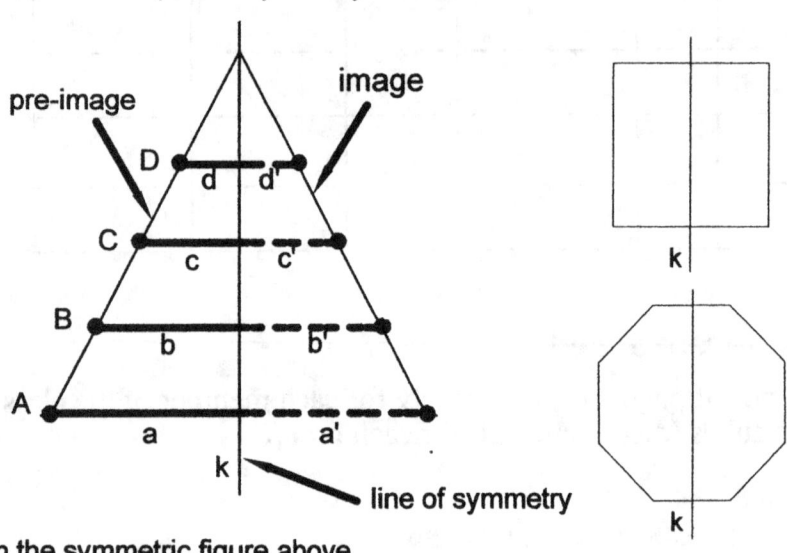

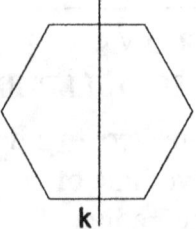

In the symmetric figure above,
k is the line of symmetry
a is equal to a'
b is equal to b'
c is equal to c'. and
d is equal to d'.
* That is, the image distance to the line of symmetry is always equal to the pre-image distance to the line of symmetry.

In all of the above figures, k is the line of symmetry.

Directions

Write in the boxes below the coordinates of the image points (from A' to K')
of the given pre-image points (from A to K).
The letters in the checked boxes will spell out the answer to the following
puzzle.

Why dos this turkey strut across the street?

A(8, 15)   A' (☐ᵛ , ☐ᵛ)   G(13, 20)   G'(☐ᵛ , ☐ᵛ)

B(12. 18)   B' (☐ᵛ , ☐ᵛ)   H(3ᵛ , 8ᵛ)   H'(☐ , ☐)

C(13, 22)   C' (☐ᵛ , ☐ᵛ)   I (9ᵛ , 3ᵛ)   I'(☐ , ☐)

D (5ᵛ , 8ᵛ)   D' (☐ , ☐)   J(11ᵛ , 5ᵛ)   J'(☐ , ☐)

E (5ᵛ , 9ᵛ)   E' (☐ , ☐)   K(14ᵛ , 0)   K'(☐ , ☐)

F(9, 14)   F' (☐ᵛ , ☐ᵛ)

Copy the numbers in the checked boxes above, write them
under the boxes as shown below, and then write in the
box the letter of the alphabet corresponding to the number.

| T | O |ᵛ| ☐ᵛ | ☐ᵛ | ☐ᵛ | ☐ᵛ | | ☐ᵛ | | ☐ᵛ | | ☐ᵛ | ☐ᵛ | | ☐ᵛ | ☐ᵛ | ☐ᵛ | ☐ᵛ | ☐ᵛ |
20 15

The

Struttin'

Turkey

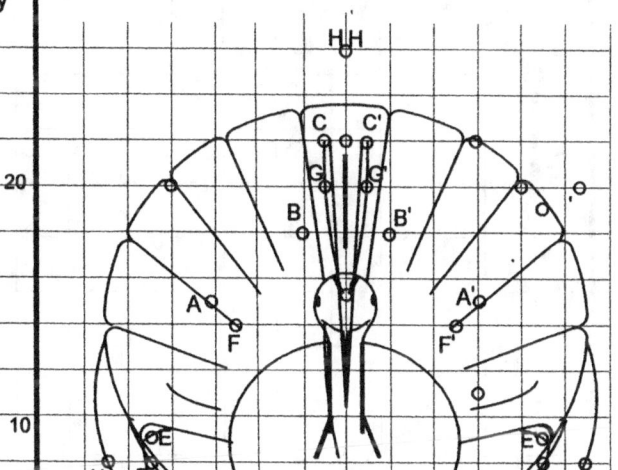

Optional enrichment exercises

ON YOUR OWN
TRY THESE SKILLBUILDERS AND RECALL ENHANCERS

1. Write on the blanks below the coordinates (ordered pairs) of the image points corresponding to the pre-image points as shown in the figure below.

A' (__ , __)   F' (__ , __)

B' (__ , __)   G' (__ , __)

C' (__ , __)   H' (__ , __)

D' (__ , __)   I' (__ , __)

E' (__ , __)   J' (__ , __)

2. Draw a line through the image points of the given pre-image points below to complete the figure.

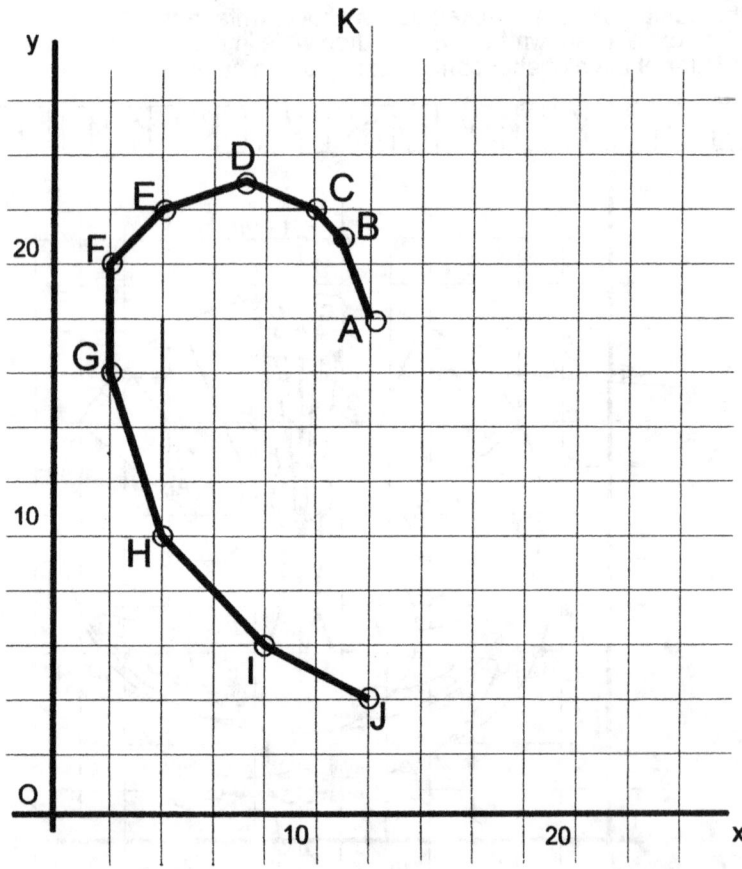

Learning goal

To construct a sphere and show its area to be equal to $4 \times \pi \times r^2$.

Directions

A sphere is not the same object as a circle. A circle is flat, whereas a sphere has depth and looks circular any way you look at it.
The accepted value for the area A, or surface area A, of a sphere is
$A = 4 \pi r^2$.

Step 1. Divide the class into groups of 4. Each group member builds a fourth or quadrant of a sphere. The 4 quadrants can be assembled to form a whole sphere. On a half-size posterboard or Manila folder, draw a circle with radius 4".

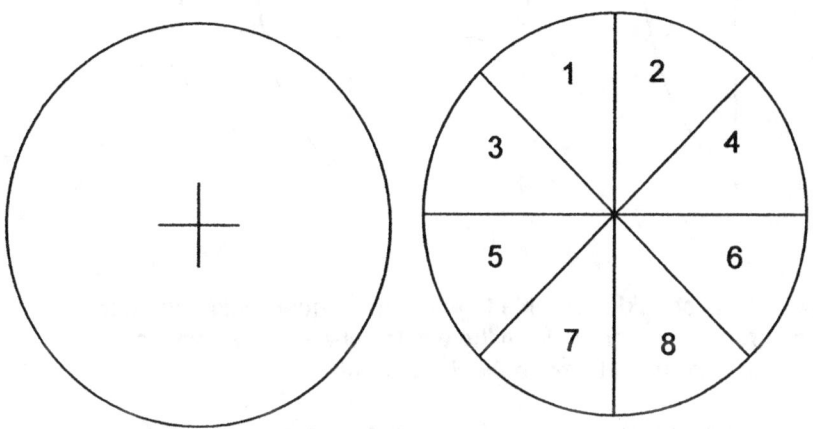

Step 2. Divide the circle into 8 sectors (divisions) as shown above at the right. Cut out the circle, then cut out the sectors, then tape or glue them together as shown below.

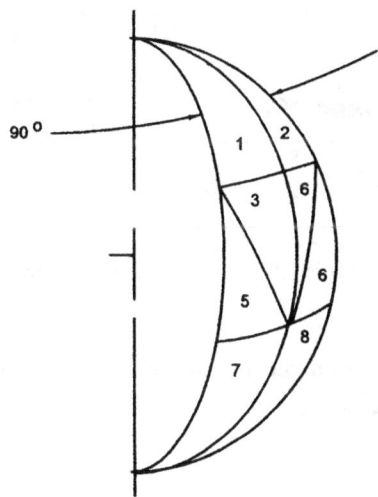

Step 3.   Draw another circle with the same 4" radius, fold it through the center
to form a 90° angle as shown at the left below. Insert into this 90°
fold the assembled quadrant at the right, and glue it to the sides of
the fold

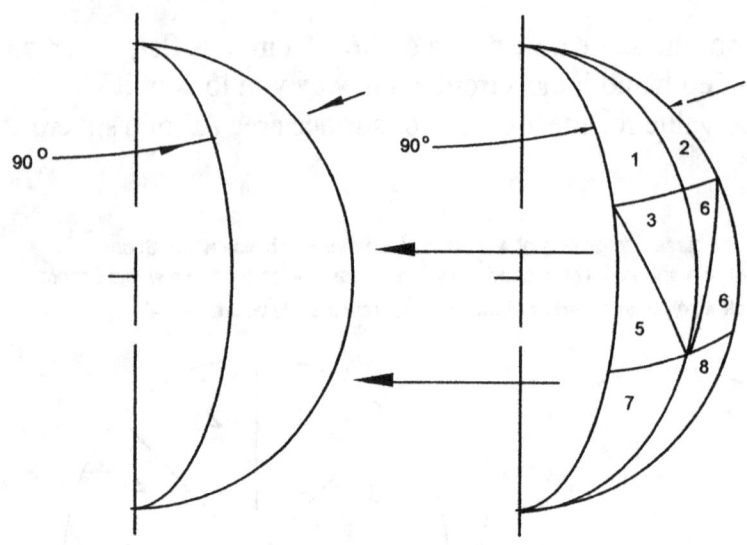

Step 4.   Each group should assemble together the 4 quadrants, one from
each group member, to form the whole sphere. Observe that
4 circles were used to make the 4 quadrants.

The area of one circle is πr^2, hence the area of the whole sphere
equals the area of 4 circles or a total of $4 \pi r^2$.

Step 5.   The group that built the neatest-looking sphere WINS.

Optional enrichment exercises

ON YOUR OWN
TRY THESE SKILLBUILDERS AND RECALL ENHANCERS

1. Find the area of a sphere whose radius is 12". Leave yur answer in terms of pi, and round
the answer to 2 decimal places.

2. If the diameter of a sphere is 16", determine its area. Leave your answer in terms of pi, and
round the answer to 2 decimal places.

Sharks play what game in the ocean?

Learning goal

To have a better understanding of ordered pairs (coordinates) on the coordinate plane, as well as the circumference and area of circles.

Suggestions for teachers

Explain again, and give more examples of, the concept of ordered pairs, how they are used to locate points on the coordinate plane. Also, differentiate the circumference of a circle from its area. Review as well how to find the area of a triangle.

What to prepare in advance

A photocopy of this page for each member of the class.

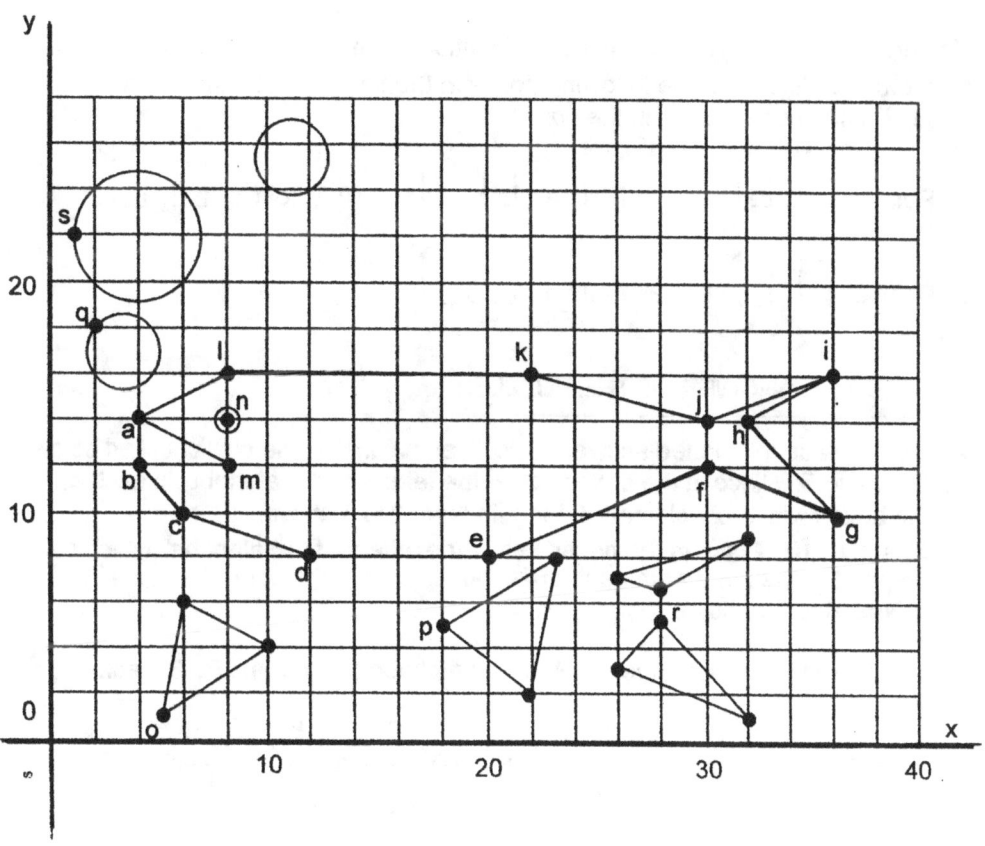

 **25**

Problem

| What kind of game do sharks play in the ocean? |

The solution to the above problem can be found in the answers to the following questions.

Circle the correct answer:

1. What do we do with food after chewing it?   return to plate   swallow
Place the letters of your answer in the boxes at the right:

2. The area of a triangle is equal to 1/2 base times ☐☐☐☐☐☑ .

3. The circular part of cars which they ride on is the ☐☐☑☐☐ .

4. The circumference of a circle is equal to the product of pi and the

☐☐☐☐☐☑☐☐ .

5. The big ride at the fair or carnival which carries people up and down in

a circle is called ☐☐☐☐☐☐ ☐☐☐☐☑ .

The ordered pair (x, y) determines the location of any point on a plane.
Find the coordinates of the following points in the figure on the preceding page: Place your answers in the boxes.

6. Point c ☑ , ☐   8. Point a ☑ , ☐   10. Point q ☐ , ☑

7. Point s ☐ , ☑   9. Point p ☐ , ☑

TO FIND THE ANSWER TO THE PUZZLE

A. No 1. Place the circled answer in No. 1 above inside the oval checked below.
B. Nos. 2 to 5. Place in the boxes below the letters of the alphabet in each of
the boxes with a check mark (✓) in Nos. 2 to 5 above.
C. Nos. 6 to 10. Place in the boxes below the letters of the alphabet which correspond to the number in each of the boxes with a check mark (✓) in
in Nos. 6 to 10 above.

Example: 1 means the first letter, A, in the alphabet; 2 means B; 3 means C,
etc.

1   2 3 4   5 6 7 8 9 10

◯   ☑☑☑   ☑☑☑☑☑☑

 **25**

Optional enrichment exercises

ON YOUR OWN
TRY THESE SKILLBUILDING EXERCISES AND RECALL ENHANCERS

1. Describe the figure formed by connecting with straight lines the points listed below: (For convenience, graphing paper may be used).

   A (10, 22)   E (10, 2)

   B (8, 14)   F (12, 10)

   C (0, 12)   G (20, 12)

   D (8, 10)   H (12, 14)

2. Determine the area of the whole figure. All measurements are in inches.

ANSWERS

LAND OF THE PILGRIMS' PRIDE

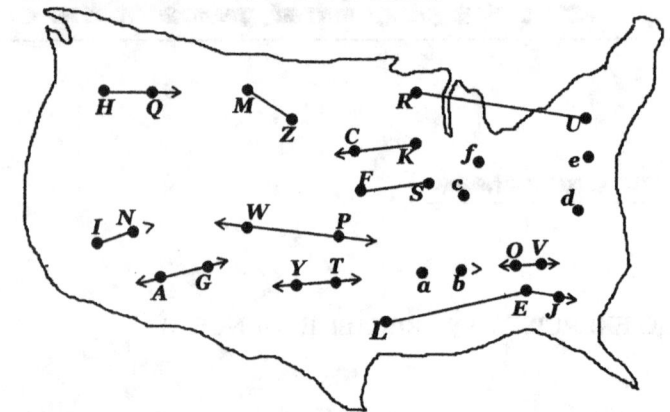

1. segment RU
2. ray EJ
3. line WP
4. line OV
5. segment LE
6. segment FS
7. line YT
8. line AG
9. segment MZ

ENRICHMENT EXERCISES

1. segment EF
2. ray KL

3. segment ST
4. ray RV
5. point L
6. plane ABCD
7. OC = 30
8. AB = 20
9. AC = 10
10. BE = 15
11. OE = 25
12. YB = 15
13. YO = 5
14. YC = 20
15. WD = 30
16. Y should be 18

RACE CARS ON THE CIRCLE SPEEDWAY

OPTIONAL ENRICHMENT EXERCISES

1. vertex   2. acute   3. triangle   4. 90   5. acute

TREK TO THE STARS

1. The name for a polygon depends upon the number of its sides. If it has 20 sides - it could be called a 20-gon, if it has 30 sides - 30 gon, if it has 40 sides - 40 gon, etc.
2. triangle AOB would be just a line   3. figure AOB would just be a line   4. just a dot
5. just a line   6. no. of sides cannot be counted   7. a circle

OPTIONAL ENRICHMENT EXERCISES

1. polygon   2. many   3. equal   4. b or d are both right   5. octagon
6. 10   7. 36

BETTER LUCK WITH HORSESHOES

OPTIONAL ENRICHMENT EXERCISES

1. scalene   2. scalene   3. No. The sum of all angles in a triangle must always be 180 degrees. The two right angles would have 180 degrees just by themselves, there would then be no thrid angle.   4. isosceles   5. 30   6. right
7. isosceles   8. obtuse

WHEEL OF PYTHAGOREAN TRIPLES

OPTIONAL ENRICHMENT EXERCISES

1. c = 2.5"   2. c = 28.28 ft   3. c = 15.62"   4. c = 12.81"   5. legs
6. 90   7. right   8. c = 32.02"

THE GREAT TRIANGLE MYSTERY

1. 55   2. 20   3. 20   4. 50   5. 30   6. 35   7. 20   8. 32   9. 35   10. 82

OPTIONAL ENRICHMENT EXERCISES

1. 180   2. 180   3. 180   4. straight

QUADRILATERAL GEO-NOPOLY

PERIMETERS:

a. 30"   b. 24"   c. The figure can be broken down into a square at the right, and a right triangle at the left.

The base of the right triangle is 2", altitude is 6", its hypotenuse (using the Pythagorean Theorem) is c = $\sqrt{40}$ = 6.32". Thus P = 7 + 6 + 9 + 6.32 = 28.32".

d. The rhombus can be broken down into 4 right triangles with legs of 5" and 3", and a hypotenuse of c = $\sqrt{34}$ = 5.83".
Thus P = 5.83 x 4 sides = 23.32".

AREAS

a. 50 sq in   b. 36 sq in   c. A = 1/2 (9 + 7) x 6 = 48 sq in
d. A = 1/2 x 10 x 6 = 30 sq in

WHAT REALLY IS A GOLDEN RECTANGLE?

Rec-tangle	W / H			Rec-tangle	W / H	
ABCD	3.5 / 2.19 = 1.6	G		YZab	3.5 / 2.19 = 1.6	G
EFGH	6 / 2.19 = 2.31	NG		cdef	4.69 / 7.5 = .63 or 7.5 / 4.69 = 1.6	G
IJKL	3.5 / 2.19 = 1.6	G		ghij	6.25/ 3.9 = 1.6	G
MNOP	3.5 / 2.19 = 1.6	G		klmn	4.69 / 7.5 = .63 or 7.5 / 4.69 = 1.6	G
QRST	6 / 2.19 = 2.31	NG		opqr	7.5 / 4.69 = 1.6	G
UVWX	3.5 / 2.19 = 1.6	G				

Part 2
Rectangle klmn, area = 35.18 sq ft
Rectangle opqr, area = 35.10 sq ft

HOW MUCH AREA DOES A TRIANGLE HAVE/

ABC: 3 sq units   JKL: 6   sq units
WXY: 1.7   MNO: 7.5
DEF: 3.46   PQR: 12
GHI: 9

N	O	N	A	G	O	N
14	15	14	1	7	15	14

OPTIONAL ENRICHMENT EXERCISES

1. 96 sq in   2. 30 sq in   2. 150 sq in

AMAZING AREA FORMULA FOR TRIANGLES

1. s = 1/2 (5 + 12 + 13) = 15

A = $\sqrt{15 \times (15 - 12) \times (15 - 5) \times (15 - 13)}$ = $\sqrt{900}$ = 30

2. s = 1/2 (8 + 6 + 10) = 12

A = $\sqrt{12 \times (12 - 8) \times (12 - 6) \times (12 - 10)}$ = $\sqrt{576}$ = 24

3. s = 1/2 (24 + 7 + 25) = 28

A = $\sqrt{28 \times (28 - 24) \times (28 - 7) \times (28 - 25)}$ = $\sqrt{7056}$ = 84

4. s = 1/2 (12 + 9 + 15) = 18

A = $\sqrt{18 \times (18 - 12) \times (18 - 9) \times (18 - 15)}$ = $\sqrt{2916}$ = 54

5. s = 1/2 (4 + 3 + 5) = 6

A = $\sqrt{6 \times (6 - 4) \times (6 - 3) \times (6 - 5)}$ = $\sqrt{36}$ = 6

6. s = 1/2 (20 + 15 + 25) = 30

A = $\sqrt{30 \times (30 - 20) \times (30 - 15) \times (30 - 25)}$ = $\sqrt{22500}$ = 150

7. s = 1/2 (24 + 32 + 40) = 48

A = $\sqrt{48 \times (48 - 24) \times (48 - 32) \times (48 - 60)}$ = $\sqrt{147456}$ = 384

8. s = 1/2 (24 + 18 + 30) = 36

A = $\sqrt{36 \times (36 - 24) \times (36 - 18) \times (36 - 30)}$ = $\sqrt{46656}$ = 216

9. M   10. U   11. L   12. A

AMAZING AREA FORMULA FOR TRIANGLES (cont'd)

OPTIONAL ENRICHMENT EXERCISES

1. $s = 1/2 (10 + 10 + 10) = 15$

$A = \sqrt{15 \times (15 - 10) \times (15 - 10) \times (15 - 10)} = \sqrt{1875} = 43.30$ sq in

2. $s = 1/2 (10 + 14 + 20) = 22$

$A = \sqrt{22 \times (22 - 10) \times (22 - 14) \times (22 - 20)} = \sqrt{4224} = 64.99$ sq in

NO OTHER TRIANGLES LIKE THESE

A 6	2	5	░	░	P 1	░
2	░	░	C 1	6	8	0
5	░	░	2	░	4	░
░	F 9	8	0	░	0	H
░	8	░	0	L 8	░	
M 4	0	2	░	G 1	1	0
░	░	░	0	░	░	

OPTIONAL ENRICHMENT EXERCISES

1. shorter leg; longer leg; shorter

2. 1.4; equal

ALWAYS 18 IN A LINE

AREAS

Rectangle A	3	Parallelogram A	7
Rectangle B	4	Parallelogram G	5
Square C	4	Rhombus D	15
Square I	1	Trapezoid H	11
Square E	9		

D 3	F 7	B 8
H 11	A 6	I 1
C 4	G 5	E 9

OPTIONAL ENRICHMENT EXERCISES

1. 144 sq in
2. 80 sq in
3. 24 sq in
4. 64 sq in
5. 120 sq in

REDISCOVERING PI

OPTIONAL ENRICHMENT EXERCISES

1. 62.86"
2. 37.71"
3. 47.14"
4. 47.14
5. 40"
6. 60"

AROUND THE HOUSE IN CIRCLES

C = 4π D = _4_	D = 4 A = _4π_	A = 100π r = _50_	A = 16π r = _8_	C = 16π D = _8_	D = 22 A = _121π_	A = 49π r = _24.5_	D = 20 C = _20π_	C = 24π D = _24_
D = 6 C = _6π_	A = 4π r = _8_	C = 6π D = _6_	D = 6 A = _9π_	D = 4 C = _4π_	D = 30 A = _225π_	A = 64π r = _32_	C = 18π D = _18_	D = 18 C = _18π_
D = 2 A = _π_	A = 9π r = _4.5_	D = 8 C = _8π_	A = 25π r = _5_	A = 36π r = _6_	C = 16π D = _8_	A = 81π r = _9_	D = 20 A = _100π_	D = 24 A = _144π_
C = 10π D = _10_	D = 12 C = _12π_	D = 8 A = _16π_	C = 12π D = _12_	D = 14 A = _49π_	D = 14 C = _14π_	D = 16 A = _64π_	C = 20π D = _20_	D = 24 C = _24π_

Picture Puzzle Solution Guide

CIRCLES FROM TRIANGLES

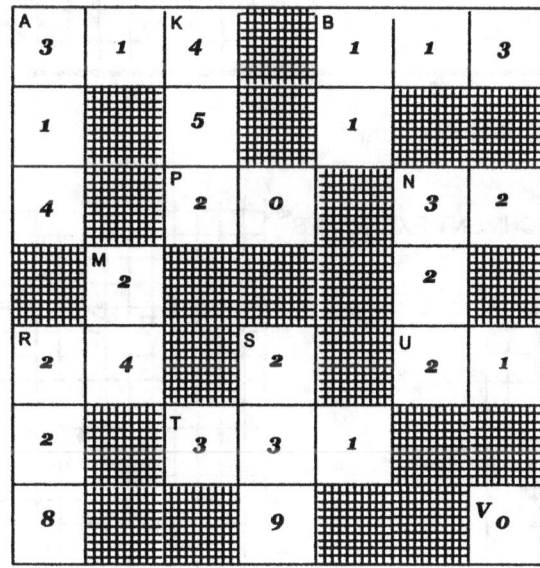

CIRCLES ON THE TARGET BOARD

OPTIONAL ENRICHMENT EXERCISES

1. π x dia.
2. π x r
3. 100 π
4. 10.96 π

A HAT FOR MAGICIANS AND CLOWNS

1. 200π
2. 120π
3. 270π

4. 15π
5. 32π
6. 64π

OPTIONAL ENRICHMENT EXERCISES
1. 12π
2. 100π

2. 96π
4. 800π

ROBIN HOOD COULD HAVE DRAWN THIS WITH HIS ARROWS

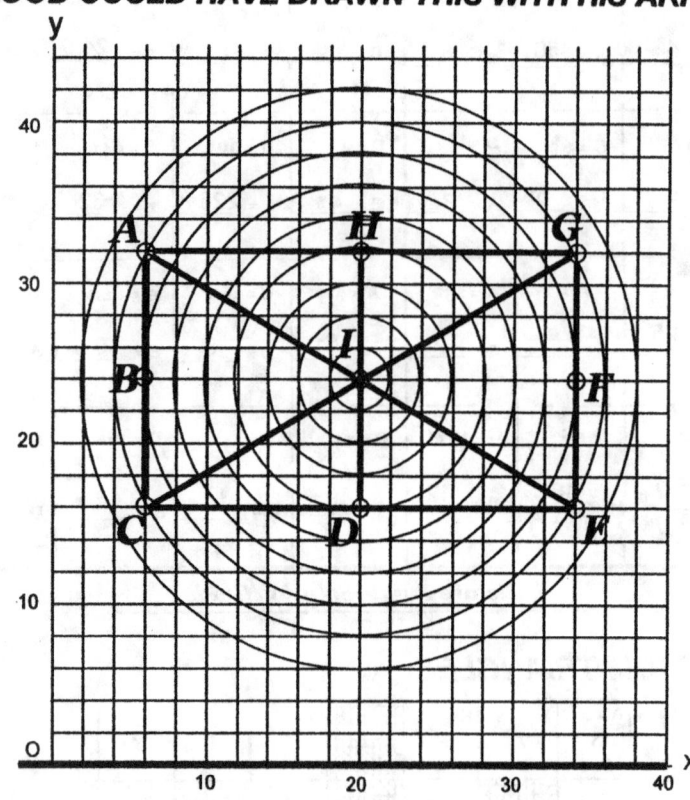

The Flag of the United Kingdom

OPTONAL ENRICHMENT EXERCISES

1, A straight line

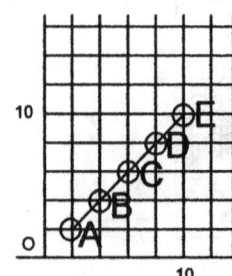

2. An open book

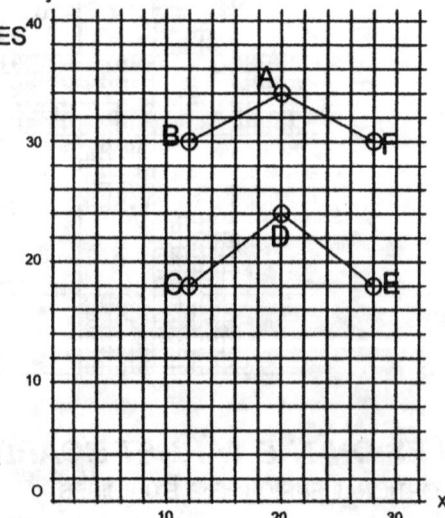

MAKE A HAT FOR 'MARDI GRAS"

OPTIONAL ENRICHMENT EXERCISES
1. 130π
2. 12π

THE LONGEST FLIGHT

1. Volume A : 72 cu ft
2. Volume B : 24 cu ft
3. Volume C : 27 cu ft
4. Volume D : 20.93 cu ft

WHY THIS TURKEY STRUTS ACROSS THE STREET

A(8, 15)	A' (20✓ , 15✓)	G(13, 20)	G' (15✓ , 20✓)
B(12. 18)	B' (16✓ , 18✓)	H (3✓ , 8✓)	H' (25 , 8)
C(13, 22)	C' (15✓ , 22✓)	I (9✓ , 3✓)	I' (20 , 3)
D (5✓ , 8✓)	D' (23 , 8)	J (11✓ , 5✓)	J' (17 , 5)
E (5✓ , 9✓)	E' (23 , 9)	K (14✓ , 0)	K' (14 , 4)
F(9, 14)	F' (19✓ , 14✓)		

T✓O P✓R✓O✓M✓E✓ H✓E✓ I✓S N✓ O✓T✓ C✓H✓ I✓C✓ K✓ E✓N✓

OPTIONAL ENRICHMENT EXERCISES

A' (12, 18)
B' (13, 21)
C' (14, 22)
D' (17, 23)
E' (20, 22)

F' (22, 20)
G'(22, 16)
H' (20, 10)
I' (16, 6)

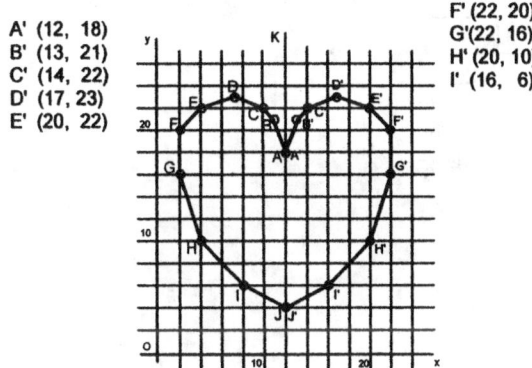

SHARKS PLAY WHAT GAME IN THE OCEAN?

1. S W A L L O W
2. H E I G H (T)
3. W (H) E E L
4. D I A M (E) T E R
5. F E R R I S   W H E E (L)
6. ((5,) 1)
7. ((1,) 22)
8. ((4,) 14)
9. (18, (5))
10. (2, (18))

S W A L L O W

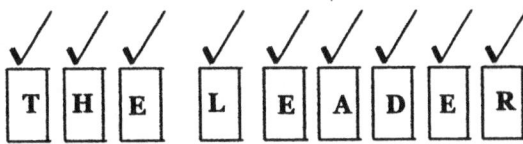

T H E L E A D E R

SHARKS PLAY WHAT GAME IN THE OCEAN? (cont'd)

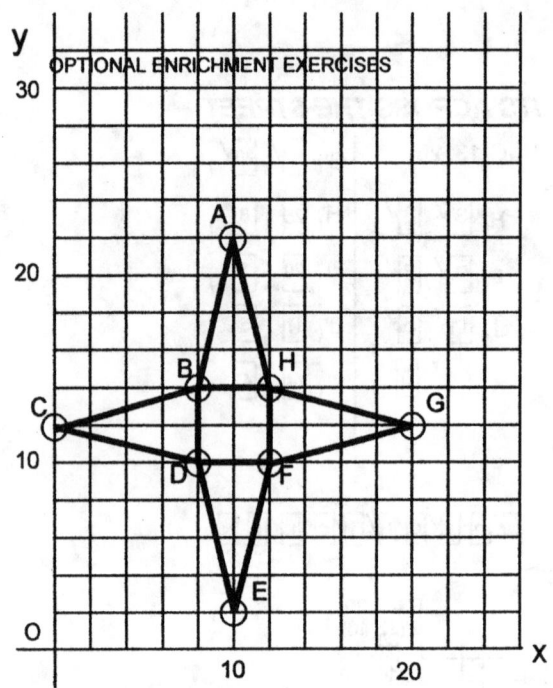

OPTIONAL ENRICHMENT EXERCISES

1. A four-pointed star

2. Area = A of square + A of four triangles
     = (4 x 4)   + 4 (1/2 x 4 x 8)
     = 80 sq in